ENGINEERING UNIT CONVERSIONS

Second Edition

Michael R. Lindeburg, P.E.

PROFESSIONAL PUBLICATIONS, INC.
Belmont, CA 94002

In the ENGINEERING REFERENCE MANUAL SERIES

Engineer-In-Training Reference Manual
    Engineering Fundamentals Quick Reference Cards
    Engineer-In-Training Sample Examinations
    Mini-Exams for the E-I-T Exam
    1001 Solved Engineering Fundamentals Problems
    E-I-T Review: A Study Guide
Civil Engineering Reference Manual
    Civil Engineering Quick Reference Cards
    Civil Engineering Sample Examination
    Civil Engineering Review Course on Cassettes
    Seismic Design of Building Structures
    Seismic Design Fast
    Timber Design for the Civil P.E. Exam
    Fundamentals of Reinforced Masonry Design
    246 Solved Structural Engineering Problems
Mechanical Engineering Reference Manual
    Mechanical Engineering Quick Reference Cards
    Mechanical Engineering Sample Examination
    101 Solved Mechanical Engineering Problems
    Mechanical Engineering Review Course on Cassettes
    Consolidated Gas Dynamics Tables
Electrical Engineering Reference Manual
    Electrical Engineering Quick Reference Cards
    Electrical Engineering Sample Examination
Chemical Engineering Reference Manual
    Chemical Engineering Quick Reference Cards
    Chemical Engineering Practice Exam Set
Land Surveyor Reference Manual
Petroleum Engineering Practice Problem Manual
Expanded Interest Tables
Engineering Law, Design Liability, and Professional Ethics
Engineering Unit Conversions

In the ENGINEERING CAREER ADVANCEMENT SERIES

How to Become a Professional Engineer
The Expert Witness Handbook—A Guide for Engineers
Getting Started as a Consulting Engineer
Intellectual Property Protection—A Guide for Engineers
E-I-T/P.E. Course Coordinator's Handbook
Becoming a Professional Engineer

ENGINEERING UNIT CONVERSIONS
Second Edition

Printed in the United States of America

ISBN: 0-912045-29-9

Professional Publications, Inc.
1250 Fifth Avenue, Belmont, CA 94002
(415) 593-9119

Current printing of this edition (last number): 6 5 4 3 2

TABLE OF CONTENTS

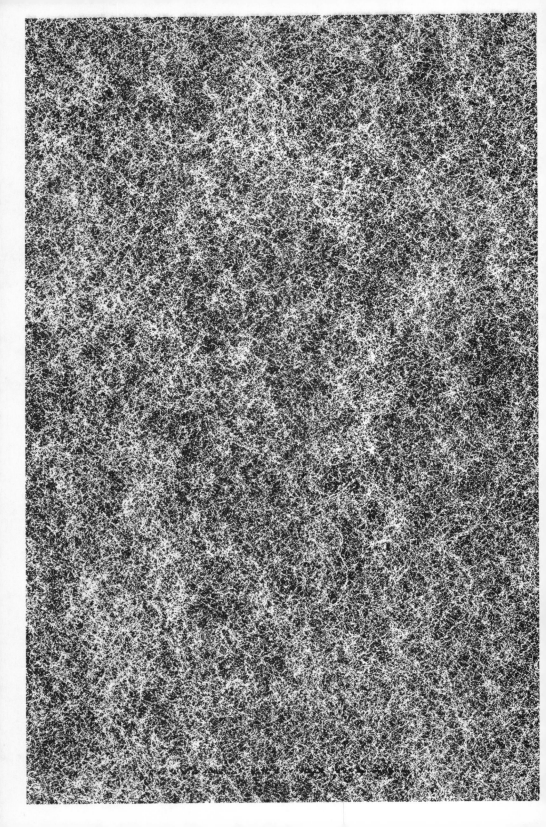

PREFACE
to the Second Edition

I can see now that the "final" edition of this book will never be written. The suggestions I have received for including additional units have shown me that the needs of the scientific and engineering communities are far more diverse than I originally anticipated.

In response to requests from readers, several hundred new conversions have been included in this edition. In addition, a thorough review of each conversion was made to ensure its accuracy. The organization of this edition is essentially unchanged, although the book has been reformatted to accommodate the changes.

Having taken the previous two years' suggestions seriously, I expect this edition will only change the nature of, not eliminate, the suggestions for improvement I receive. I like it that way. Keep those suggestions coming.

Michael R. Lindeburg, P.E.
Belmont, CA
October 1990

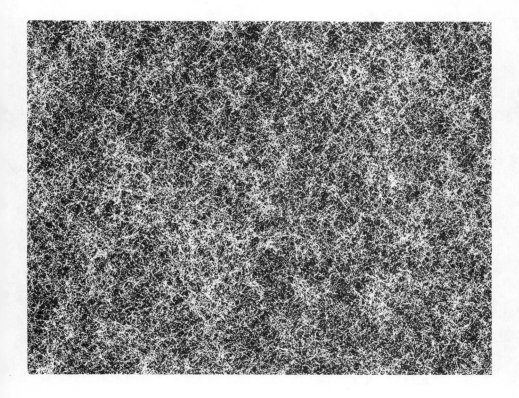

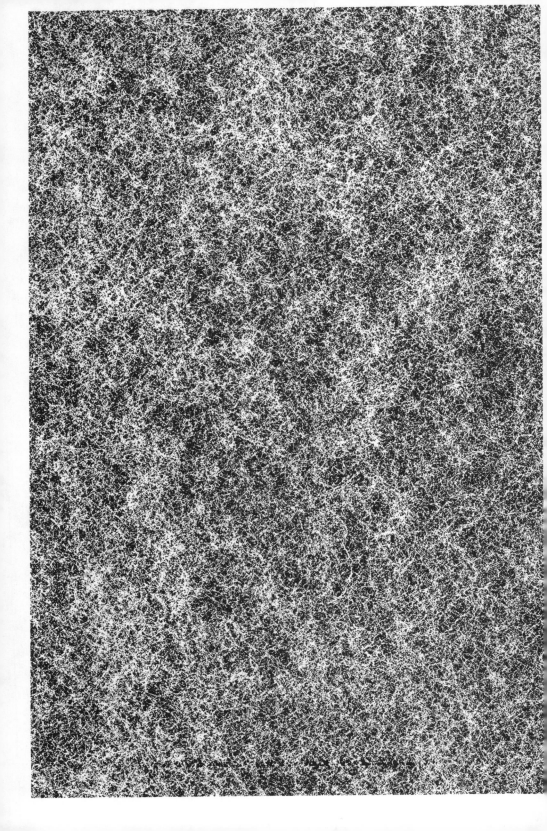

PREFACE
to the First Edition

Engineering, by its very nature, is a blend of other disciplines. Engineering can look to mathematics, physics, chemistry, and construction trades, among others, for its roots. Furthermore, engineering is not limited geographically to one country. There are engineers working in every country in the world.

It comes as no surprise, then, that engineers are often faced with the need to work with units unfamiliar to them. In some instances, engineers choose to convert foreign data into forms they are familiar with. In other cases, engineers need to convert their familiar data into foreign forms.

Engineering Unit Conversions was written to assist engineers in making such conversions. Although some basic conversions are available at the touch of a calculator button, most conversions needed by engineers are not. In particular, conversions of compound units (e.g., converting milligrams per liter per day into pounds per million gallons per day) must often be developed from the basic units. *Engineering Unit Conversions* eliminates that need.

This book is a great time-saver and a perfect addition to any engineer's or engineering student's library. *Engineering Unit Conversions* is recommended for anyone who works numerous problems from a variety of engineering disciplines. Engineering students and engineers preparing for their licensing exams, in particular, will find this book timeless and invaluable.

Michael R. Lindeburg, P.E.
Belmont, CA
July 1988

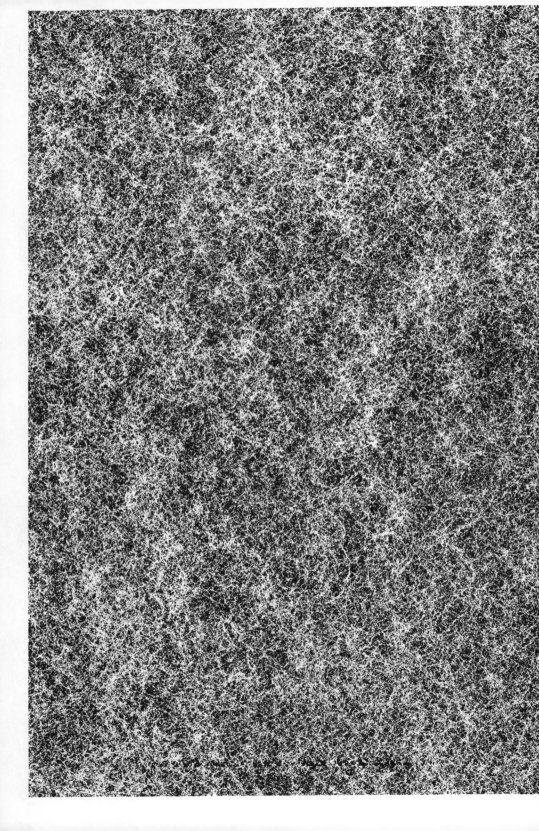

ACKNOWLEDGMENTS
for the Second Edition

For the second edition of this book, my hat is off to John Grandy, Michel Bilello, and Katherine Thompson for verifying the conversions and reconciling the differences found between sources. As usual, Shelley Arenson came to the rescue by writing page-formatting macros. Those macros greatly simplified the production process after Mary Christensson typeset the changes. Kurt Stephan performed the quality-assurance check by proofing the final work.

Due to the nature of the material in this book, these individuals can personally attest to the fact that going cross-eyed takes only about 4.7 minutes, on the average.

And, of course, my thanks to all of the people who offered their suggestions. You all helped me to produce a better book.

Michael R. Lindeburg, P.E.
Belmont, CA
October 1990

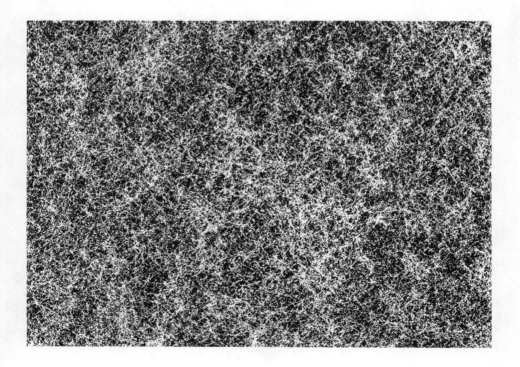

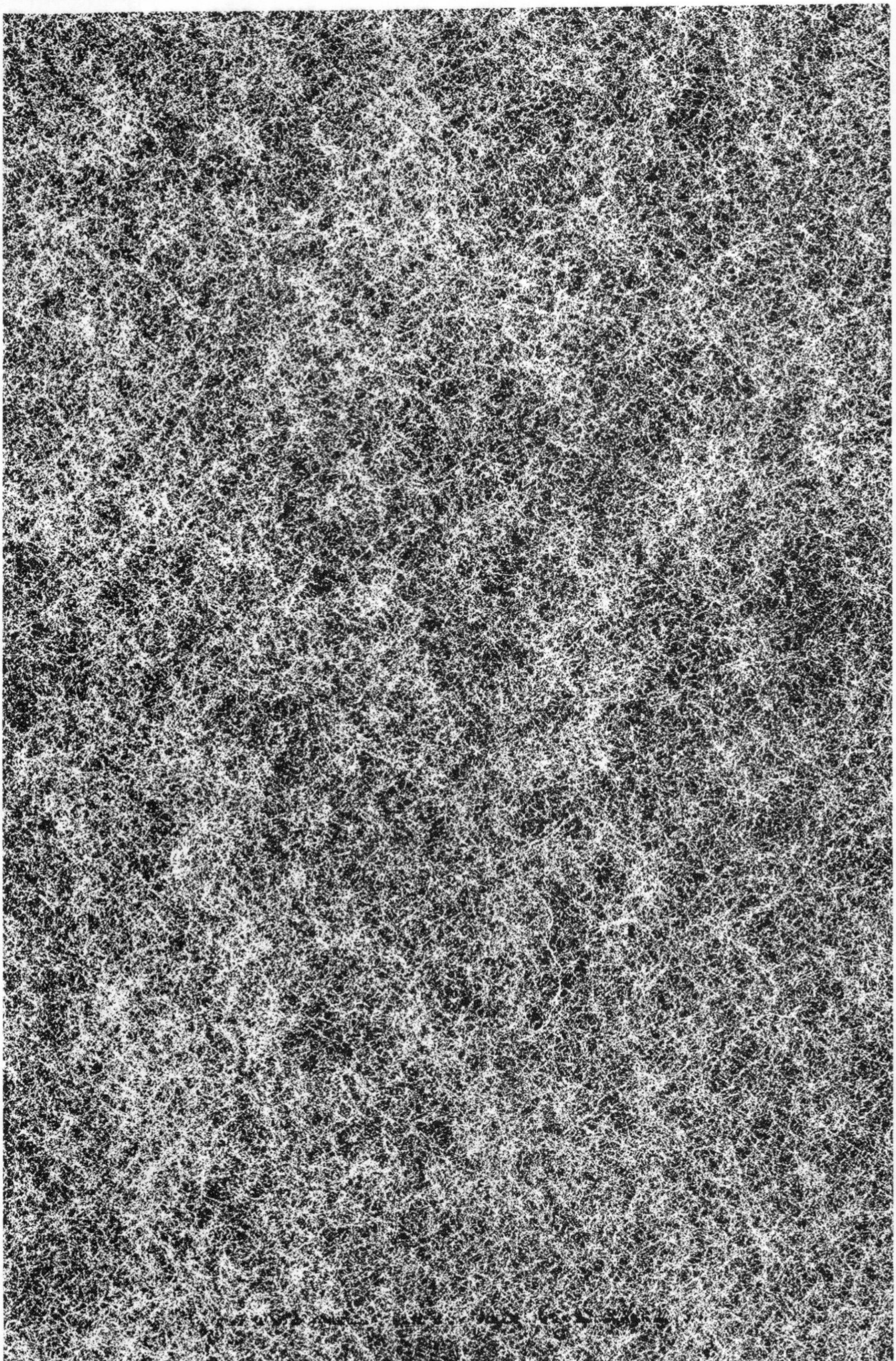

ACKNOWLEDGMENTS
for the First Edition

I tip my hat to Mr. Darrin Wagner for devoting many months of his life to compiling the information contained in this book. Darrin did the lion's share of research, compilation, and cross-checking on this book, as well as keyboarding a portion of the final manuscript.

Sylvia Osias of Professional Publication's production department performed the typesetting. Thank you, Sylvia, for never growing weary as changes, additions, and reorganizations were made to the manuscript.

Thank you Cindy Arnold, of Professional Publication's art department, for developing the cover design and pasting up the book. It is unfortunate that readers can't see the many preliminary cover designs that weren't selected—they were excellent also.

Thank you Tom Whipple, of Technical Editing Services, for proofreading the typeset galleys against the original manuscript, for making great suggestions on style and organization, and for helping Professional Publications stay on schedule.

And, none of this would have been possible without Lisa Rominger, production department supervisor, for developing the page format, overseeing the project, and keeping it on schedule. If you ever became frustrated with me as I tried to squeeze "just one more" conversion into the book, you never showed it.

Michael R. Lindeburg, P.E.
Belmont, CA
July 1988

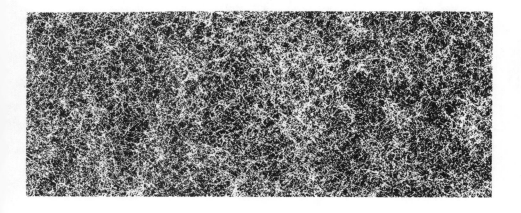

Notice to Examinees

Do not copy, memorize, or distribute problems from the Fundamentals of Engineering (FE) Examination or the Principles and Practice of Engineering (P&P) Examination. These acts are considered to be exam subversion.

The FE and P&P examinations are copyrighted by the National Council of Examiners for Engineering and Surveying. Copying and reproducing FE and P&P exam problems for commercial purposes is a violation of federal copyright law. Reporting examination problems to other examinees invalidates the examination process and threatens the health and welfare of the public.

INTRODUCTION

A few words about how this book is arranged will make it easier to use. Familiarity with the following guidelines will assist you in locating the conversion you need.

Conversions in this book are listed alphabetically. Each unit for which a conversion factor is given appears once in the first (MULTIPLY BY) column of the table. The unit is omitted from the first column for subsequent conversions of that unit.

In addition to being listed alphabetically, units by themselves are listed before their compound units (i.e., dynes comes before dyne-cm). Specifically, single units are given first (e.g., feet), units divided by another unit come next (e.g., feet/sec), and units multiplied by another unit are last (e.g., foot-lbf). Units in the third (TO OBTAIN) column of the table are alphabetized in the same manner.

During the planning stage for this book, it was difficult to come up with a standard presentation of the units in a form that all engineers would recognize and that would lend itself to an aesthetic page layout. Abbreviations and symbols commonly used by engineers were too brief and left too much room for interpretation. However, space limitations did not allow for spelling out the name of each unit in its entirety, so a compromise was made. Units are spelled out up to a slash (representing "per") or hyphen (for compound units), at which point, common abbreviations are used.

For example, the metric unit of power is listed as "watts." Watts per hour is listed as "watts/hr" (watts is the sole numerator). The compound unit watt-hours per kilogram is listed as "watt-hr/kg."

The convention of spelling out units means that some units will not appear in their most common forms. Pounds per square inch is found under "pounds/in^2," and not under the commonly used abbreviation "psi." Cubic feet (most commonly represented as ft^3) is in the "C" section listed under "cubic feet," and not in the "F" section under "ft^3" or "feet cubed." However, ft^3 is the abbreviation used when cubic feet appears after a hyphen in a compound unit (e.g., poise-ft^3/lbm) or after a slash (e.g., gallons/ft^3). Similarly, square feet is found in the "S" section.

Abbreviations used in this book and their meanings are listed in the following table.

abbrev.	meaning	abbrev.	meaning
accel.	acceleration	IST	international steam tables
amp	ampere	kg	kilogram
amu	atomic mass unit	KS	Kansas
atm	atmosphere	ℓ or l	liter
avoir	avoirdupois	lbf	pound (force)
AZ	Arizona	lbm	pound (mass)
BeV	billion electron volts	m	meter
BTU	British thermal unit	MeV	million electron volts
C	Centigrade	MGD	million gallons per day
CA	California	mi	mile
cal	calorie	min	minute
CHU	caloric heat unit	ml	milliliter
cm	centimeter	mm	millimeter
CO	Colorado	MT	Montana
F	Fahrenheit	ND	North Dakota
ft	feet	NE	Nebraska
g	gram	NM	New Mexico
g	gravity	NV	Nevada
gal	gallon	OR	Oregon
GeV	giga electron volts	ppm	parts per million
gpd	gallons per day	SD	South Dakota
H_2O	water	sec	second
Hg	mercury	thermo	thermodynamics
hr	hour	U.S.	United States
ID	Idaho	UT	Utah
in	inch	wt	weight
int'l	international	yd	yard

A Note for Engineers in the United States

U.S. engineers may, at first, find it confusing when they see how many different types of barrels and tons (and many other units listed in this book) there are. When a U.S. engineer says "barrel," 42 gallons is implied. The normal quantities used in the U.S. are:

barrel	42 gallons
BTU	traditional (thermochemical) value
calories	thermochemical
gallons	U.S., liquid
horsepower	U.S. or mechanical
ounces	avoirdupois
pints	U.S., liquid
pounds	avoirdupois
quarts	U.S., liquid
tons	short tons of 2000 pounds

PROFESSIONAL PUBLICATIONS, INC. ● Belmont, CA

MULTIPLY	BY	TO OBTAIN
abamperes	10.	amperes
	1.03627×10^{-4}	faradays/sec
	10,000.	milliamperes
	2.99793×10^{10}	statamperes
abamperes/cm^2	10.	amperes/cm^2
	64.52	amperes/in^2
	100,000.	amperes/m^2
	2.998×10^{10}	statamperes/cm^2
abampere-turns	10.	ampere-turns
	12.566	gilberts
abampere-turns/cm	10.	ampere-turns/cm
	25.4	ampere-turns/in
	1000.	ampere-turns/m
	12.566	gauss (also oersteds)
	12.566	gilberts/cm
abcoulombs	0.0027778	ampere-hr
	10.	coulombs
	6.24126×10^{19}	electronic charges
	8.3668×10^{6}	electrostatic ft-lbf-sec
	1.0364×10^{-4}	faradays
	2.998×10^{10}	statcoulombs
abcoulombs/cm^2	10.	coulombs/cm^2
	64.52	coulombs/in^2
	100,000.	coulombs/m^2
abfarads	1.0×10^{9}	farads
	1.0×10^{15}	microfarads
	1.0×10^{21}	micromicrofarads
	1.0×10^{21}	picofarads
	8.98755×10^{20}	statfarads
abhenrys	1.0×10^{-9}	henrys
	0.001	microhenrys
	1.0×10^{-6}	millihenrys
	1.11265×10^{-21}	stathenrys
abmhos	1.0×10^{9}	mhos
	1000.	micromhos
	1.0×10^{6}	millimhos
	1.0×10^{9}	siemens
	8.988×10^{20}	statmhos
abohms	1.0×10^{-15}	megohms
	0.001	microhms
	1.0×10^{-9}	ohms (absolute)
	1.1127×10^{-21}	statohms

PROFESSIONAL PUBLICATIONS, INC. • Belmont, CA

MULTIPLY	BY	TO OBTAIN
abvolts	0.01	microvolts
	1.0×10^{-5}	millivolts
	3.3356×10^{-11}	statvolts
	1.0×10^{-8}	volts (absolute)
	9.9967×10^{-9}	volts (international)
acres	40.4687	ares
	0.404687	hectares
	5.645×10^{-3}	labors
	4.	roods
	0.001563	sections
	4.04687×10^7	square centimeters
	10.	square chains (Gunter's or surveyor's)
	43,560.	square feet
	6.27266×10^6	square inches
	0.00404687	square kilometers
	100,000.	square links (Gunter's or surveyor's)
	4046.87	square meters
	0.0015625	square miles (statute)
	160.	square rods (also perches or rods)
	4840.	square yards
	4.34028×10^{-5}	townships
acre-ft	7757.8	barrels (42 U.S. gallons)
	43,560.	cubic feet
	1233.5	cubic meters
	1613.3	cubic yards
	325,851.	gallons (U.S., liquid)
	1.876×10^{-2}	square mile-inch
	43.56	thousands of cubic feet
acre-ft/day	0.50417	cubic ft/sec
acre-ft/mile^2	0.01876	inches of runoff
acre-inches/hr	1.008	cubic ft/sec
amperes	0.1	abamperes
	1.	coulombs/sec
	0.01	edisons
	1.0363×10^{-5}	faradays/sec
	1000.	milliamperes
	2.99793×10^9	statamperes
amperes/cm^2	0.1	abamperes/cm^2
	6.4516	amperes/in^2
	10,000.	amperes/m^2
	2.998×10^9	statamperes/cm^2

MULTIPLY	BY	TO OBTAIN
amperes/in^2	0.0155	abamperes/cm^2
	0.155	amperes/cm^2
	1550.	amperes/m^2
	4.647×10^8	statamperes/cm^2
amperes/m^2	1.0×10^{-4}	amperes/cm^2
	6.4516×10^{-4}	amperes/in^2
	2.998×10^5	statamperes/cm^2
ampere-hr	360.	abcoulombs
	3600.	coulombs
	2.2612×10^{22}	electronic charges
	3.0120×10^9	electrostatic ft-lbf-sec
	0.037306	faradays
	1.0793×10^{13}	statcoulombs
ampere-turns	0.1	abampere-turns
	1.2566	gilberts
ampere-turns/cm	0.1	abampere-turns/cm
	2.54	ampere-turns/in
	100.	ampere-turns/m
	125,660.	gammas
	1.2566	gauss
	1.2566	gilberts/cm
	1.2566	lines/cm^2
	8.1076	lines/in^2
	1.2566×10^{-4}	teslas
ampere-turns/in	0.03937	abampere-turns/cm
	0.3937	ampere-turns/cm
	39.37	ampere-turns/m
	49,474.	gammas
	0.49474	gauss
	0.49474	gilberts/cm
	0.49474	lines/cm^2
	3.1920	lines/in^2
	4.947×10^{-5}	teslas
ampere-turns/m	0.001	abampere-turns/cm
	0.01	ampere-turns/cm
	0.0254	ampere-turns/in
	1256.6	gammas
	0.012566	gauss
	0.012566	gilberts/cm
	0.012566	lines/cm^2
	0.081076	lines/in^2
	1.2566×10^{-6}	teslas

MULTIPLY	BY	TO OBTAIN
angstroms	1.0×10^{-8}	centimeters
	3.28084×10^{-10}	feet
	1.0×10^{5}	fermis
	3.9370×10^{-9}	inches
	1.0×10^{-13}	kilometers
	1.0×10^{-10}	meters
	1.0×10^{-4}	micrometers
	100.	micromicrons
	1.0×10^{-4}	microns
	6.214×10^{-14}	miles
	1.0×10^{-7}	millimeters
	0.1	millimicrons
	3.9370×10^{-6}	mils
	0.1	nanometers
	100.	stigmas
	1.	tenthmeters
	1.553×10^{-4}	wavelengths red line cadmium
	997.984	X-units
	1.0×10^{5}	yukawas
apostilbs	0.029572	candles/ft^2
	2.0536×10^{-4}	candles/in^2
	0.31831	candles/m^2
	0.092902	foot-lamberts
	1.0×10^{-4}	lamberts
	3.1831×10^{-5}	lumens/cm^2-sterad
	0.029572	lumens/ft^2-sterad
	0.1	millilamberts
	3.1831×10^{-5}	stilbs
ares	0.024710	acres
	100.	centares
	0.01	hectares
	0.098842	roods
	1.0×10^{6}	square centimeters
	0.247105	square chains (Gunter's or surveyor's)
	1076.4	square feet
	155,000.	square inches
	1.0×10^{-4}	square kilometers
	2471.	square links (Gunter's or surveyor's)
	100.	square meters
	3.861×10^{-5}	square miles
	3.95367	square rods (also square perches)
	119.6	square yards
	1.07250×10^{-6}	townships

MULTIPLY	BY	TO OBTAIN
assay tons	145.833	carats
	16.461	drams (avoir)
	7.5019	drams (troy)
	450.1	grains
	29.1667	grams
	0.029167	kilograms
	1.0288	ounces (avoir)
	0.93773	ounces (troy)
	18.755	pennyweights
	22.506	scruples
astronomical units	4.908×10^{11}	feet
	1.496×10^{8}	kilometers
	1.581×10^{-5}	light years
	1.496×10^{11}	meters
	9.296×10^{7}	miles (statute)
	4.848×10^{-6}	parsecs
	1.636×10^{11}	yards
atmospheres	1.0332	atmospheres (metric)
	1.01325	bars
	76.	centimeters Hg at 0°C
	1033.3	centimeters H_2O at 4°C
	1.01325×10^{6}	dynes/cm^2
	33.900	feet H_2O at 39.1°F
	1033.2	grams/cm^2
	29.921	inches Hg at 32°F
	406.79	inches H_2O at 39.1°F
	1.0332	kilograms/cm^2
	10,332.	kilograms/m^2
	0.010332	kilograms/mm^2
	760.	millimeters Hg at 0°C
	235.13	ounces/in^2
	101,325.	pascals
	2116.2	pounds/ft^2
	14.696	pounds/in^2
	0.94474	tons (long)/ft^2
	0.006561	tons (long)/in^2
	1.0581	tons (short)/ft^2
	0.007348	tons (short)/in^2
	760.0	torrs
atmospheres (metric)	0.9678	atmospheres
	0.980665	bars
	28.96	inches Hg at 32°F
	1.	kilograms (force)/cm^2
	14.22	pounds/in^2

PROFESSIONAL PUBLICATIONS, INC. ● Belmont, CA

MULTIPLY	BY	TO OBTAIN
atomic mass units (amu)	9.316×10^8	electron volts (nuclear)
	1.661×10^{-24}	grams
	1.492×10^{-10}	joules (nuclear)
	1.661×10^{-27}	kg
	931.5	MeV
avograms	1.66036×10^{-24}	grams

PROFESSIONAL PUBLICATIONS, INC. ● Belmont, CA

MULTIPLY	BY	TO OBTAIN
bars	0.986923	atmospheres
	75.006	centimeters Hg at 0°C
	1019.7	centimeters H_2O at 4°C
	1.0×10^6	dynes/cm^2
	33.457	feet H_2O at 39°F
	1019.7	grams/cm^2
	29.530	inches Hg at 32°F
	401.47	inches H_2O at 39°F
	1.0197	kilograms/cm^2
	10,197.	kilograms/m^2
	0.010197	kilograms/mm^2
	1000.	millibars
	10,197.	millimeters H_2O at 0°C
	100.	kPa
	232.1	ounces/in^2
	1.0×10^5	pascals
	2088.5	pounds/ft^2
	14.504	pounds/in^2
	0.93238	tons (long)/ft^2
	0.0064748	tons (long)/in^2
	1.0443	tons (short)/ft^2
	0.0072519	tons (short)/in^2
barleycorns	0.84667	centimeters
	48.	douziemes
	0.027778	feet
	0.33333	inches
	4.	lines
barns	1.6×10^{-25}	square inches
	1.0×10^{-28}	square meters
	1.0×10^{-22}	square millimeters
barrels (British, dry)	5.780	cubic feet
	0.1637	cubic meters
	36.	gallons (British)
barrels (U.S., dry)	4.083	cubic feet
	0.11563	cubic meters

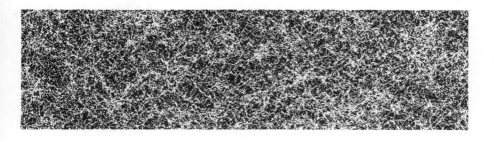

PROFESSIONAL PUBLICATIONS, INC. ● Belmont, CA

MULTIPLY	BY	TO OBTAIN
barrels (31.5 U.S. gallons)	119,241.	cubic centimeters
	4.2110	cubic feet
	7276.5	cubic inches
	0.119241	cubic meters
	0.15596	cubic yards
	3.5	firkins
	31.5	gallons (U.S., liquid)
	1008.	gills
	0.5	hogsheads
	0.119241	kiloliters
	119.241	liters
	4032.	ounces (U.S., liquid)
	252.	pints (U.S., liquid)
	126.	quarts
barrels (31.5 U.S. gallons)/day	1.3125	gallons (U.S., liquid)/hr
	0.021875	gallons (U.S., liquid)/min
	3.646×10^{-4}	gallons (U.S., liquid)/sec
barrels (42 U.S. gallons)	5.615	cubic feet
	9702.	cubic inches
	0.2080	cubic yards
	42.	gallons (U.S., liquid)
	1.	oil barrels
	336.	pints (U.S., liquid)
	168.	quarts (U.S., liquid)
barrels (42 U.S. gallons)/day	0.003899	cubic feet/min
	6.498×10^{-5}	cubic feet/sec
	1.75	gallons (U.S., liquid)/hr
	0.029167	gallons (U.S., liquid)/min
	4.861×10^{-4}	gallons (U.S., liquid)/sec
barrels (42 U.S. gallons)/hr	0.093576	cubic feet/min
	0.001560	cubic feet/sec
	2.695	cubic inches/sec
	0.7	gallons (U.S., liquid)/min
barrels (42 U.S. gallons)/min	5.615	cubic feet/min
	0.09358	cubic feet/sec
baryes	1.0×10^{-6}	bars
	1.	dynes/cm^2
becquerels	2.703×10^{-11}	curies
	1.	disintegrations/sec
bel	10.	decibels
BeV	1.0×10^{9}	electron volts

PROFESSIONAL PUBLICATIONS, INC. • Belmont, CA

MULTIPLY	BY	TO OBTAIN
board-ft	0.0052083	cord-ft
	6.5104×10^{-4}	cords
	2359.7	cubic centimeters
	0.083333	cubic feet
	144.	cubic inches
	0.0023597	cubic meters
	0.0030864	cubic yards
	0.0023597	steres
bolts	32.	ells
	120.	feet
	36.576	meters
	0.33333	skeins
	40.	yards
bougie decimales	1.	candles (int'l)
	0.104167	carcel units
	1.	English sperm candles
	1.11111	hefner units
	1.	lumens/sterad
	1.	pentane candles
BTU (traditional)	252.	calories (IST)
	252.2	calories (thermo)
	10,410.	cubic centimeter-atm
	0.3676	cubic foot-atm
	1.0548×10^{10}	dyne-cm
	1.0548×10^{10}	ergs
	25,040.	foot-poundals
	778.17	foot-lbf
	252.0	gram-cal
	1.076×10^{7}	gram-cm
	3.929×10^{-4}	horsepower-hr (U.S.)
	9339.	inch-lbf
	1054.8	joules
	0.2520	kilogram-cal
	107.6	kilogram-m
	2.930×10^{-4}	kilowatt-hr
	10.41	liter-atm
	10,548.	megalergs
	0.5555	pound-cal
	0.5555	pound-CHU
	1.0×10^{-5}	therms
	0.293	watt-hr
	1054.8	watt-sec
BTU (IST)	0.999229	BTU (mean)
	1.00067	BTU (thermo)
	1055.06	joules

MULTIPLY	BY	TO OBTAIN
BTU (mean)	1.00077 1.00144 1055.87	BTU (IST) BTU (thermo) joules
BTU (thermo)	0.999331 0.998560 1054.35	BTU (IST) BTU (mean) joules
BTU ($39°F$)	1059.67	joules
BTU ($59°F$)	1054.80	joules
BTU ($60°F$)	1054.68	joules
BTU/ft^2	2.713	kilogram-cal/m^2
BTU/ft^2-day	0.04167 0.01130 3.139×10^{-6} 1.314×10^{-5}	BTU/ft^2-hr gram-cal/cm^2-hr gram-cal/cm^2-sec watts/cm^2
BTU/ft^2-day-$°F$	0.04167 0.02034 5.651×10^{-6} 2.366×10^{-5}	BTU/ft^2-hr-$°F$ gram-cal/cm^2-hr-$°C$ gram-cal/cm^2-sec-$°C$ watts/cm^2-cm-$°C$
BTU/ft^2-hr	24. 0.2712 7.535×10^{-5} 3.155×10^{-4}	BTU/ft^2-day gram-cal/hr-ft^2 gram-cal/sec-cm^2 watts/cm^2
BTU/ft^2-hr-$°F$	24. 0.4882 1.356×10^{-4} 3.94×10^{-4} 4.882 5.682×10^{-4} 5.682 0.0011308 0.0020354	BTU/ft^2-day-$°F$ gram-cal/cm^2-hr-$°C$ gram-cal/cm^2-sec-$°C$ horsepower/ft^2-$°F$ kilogram-cal/hr-m^2-$°C$ watts/cm^2-$°C$ watts/m^2-$°C$ watts/in^2-$°C$ watts/in^2-$°F$
BTU/ft^2-min	0.0239 0.0176 0.122	horsepower (U.S.)/ft^2 kilowatts/ft^2 watts/in^2
BTU/ft^3	0.0039 8.899 3.7234×10^4	kilogram-cal/ℓ kilogram-cal/m^3 joules/m^3

PROFESSIONAL PUBLICATIONS, INC. ● Belmont, CA

MULTIPLY	BY	TO OBTAIN
BTU/ft^3-°F	6.71×10^4	joules/m^3-°C
BTU/hr	0.21616	foot-lbf/sec
	3.929×10^{-4}	horsepower (U.S.)
	2.930×10^{-4}	kilowatts
	8.333×10^{-5}	tons (refrigeration)
	0.2931	watts
BTU/lbm	0.555555	cal/g
	2326.	joules/kg
	0.5555	kilocalories/kg
	0.2930	watt-hr/lbm
BTU/lbm-°F	1.	gram-cal/gram-°C
	4186.8	joules/kg-°C
BTU/min	0.0166667	BTU/sec
	1.7584×10^8	dyne-cm/sec
	1.7584×10^8	ergs/sec
	46,685.	foot-lbf/hr
	778.17	foot-lbf/min
	12.969	foot-lbf/sec
	179,347.	gram-cm/sec
	0.02389	horsepower (metric)
	0.023575	horsepower (U.S.)
	17.576	joules/sec
	0.2519	kilogram-cal/min
	0.0041990	kilogram-cal/sec
	1.792	kilogram-m/sec
	0.017576	kilowatts
	11,750.	lumens
	0.005	tons (refrigeration)
	17.576	watts
BTU/sec	60.	BTU/min
	1.0548×10^{10}	dyne-cm/sec
	1.0548×10^{10}	ergs/sec
	2.800×10^6	foot-lbf/hr
	46,676.	foot-lbf/min
	778.17	foot-lbf/sec
	252.2	gram-cal/sec
	1.0755×10^7	gram-cm/sec
	1.434	horsepower (metric)
	1.4145	horsepower (U.S.)
	1055.	joules/sec
	15.12	kilogram-cal/min
	0.2520	kilogram-cal/sec
	107.55	kilogram-m/sec
	1.055	kilowatts
	705,030.	lumens
	1054.8	watts

PROFESSIONAL PUBLICATIONS, INC. • Belmont, CA

MULTIPLY	BY	TO OBTAIN
BTU-ft/ft^2-hr-°F	288.	BTU-in/ft^2-day-°F
	173,000.	ergs-cm/sec-cm^2-°C
	14.88	gram-cal-cm/cm^2-hr-°C
	0.004134	gram-cal-cm/cm^2-sec-°C
	1.73	joules/sec-m-°C
	0.01731	watts-cm/cm^2-°C
BTU-in/ft^2-day-°F	0.00347	BTU-ft/ft^2-hr-°F
	0.04167	BTU-in/ft^2-hr-°F
	0.05167	gram-cal-cm/cm^2-hr-°C
	1.435×10^{-5}	gram-cal-cm/cm^2-sec-°C
	6.009×10^{-5}	watts-cm/cm^2-°C
BTU-in/ft^2-hr-°F	24.	BTU-in/ft^2-day-°F
	1.241	gram-cal-cm/cm^2-hr-°C
	3.447×10^{-4}	gram-cal-cm/cm^2-sec-°C
	518.9	joules/hr-m-°C
	3.447×10^{-7}	kilocalories-cm/sec-cm^2-°C
	0.001441	watts-cm/cm^2-°C
buckets (British, dry)	4.	gallons (British)
bushels (British)	1.0321	bushels (U.S.)
	1.2844	cubic feet
	2219.4	cubic inches
	0.036369	cubic meters
	8.	gallons (British, dry)
	36.369	liters
bushels (U.S.)	0.96894	bushels (British)
	0.027778	chaldrons
	35,239.	cubic centimeters
	1.2445	cubic feet
	2150.4	cubic inches
	0.035239	cubic meters
	0.046091	cubic yards
	7.7515	gallons (British, dry)
	8.	gallons (U.S., dry)
	9.3092	gallons (U.S., liquid)
	35.239	liters
	4.	pecks
	64.	pints (U.S., dry)
	32.	quarts (U.S., dry)
	37.237	quarts (U.S., liquid)
	0.035239	steres

PROFESSIONAL PUBLICATIONS, INC. ● Belmont, CA

MULTIPLY	BY	TO OBTAIN
butts (British)	4.	barrels (31.5 U.S. gallons)
	126.	gallons (U.S., liquid)
	4032.	gills
	2.	hogsheads
	1008.	pints (U.S., liquid)
	0.9995	pipes
	504.	quarts (U.S., liquid)

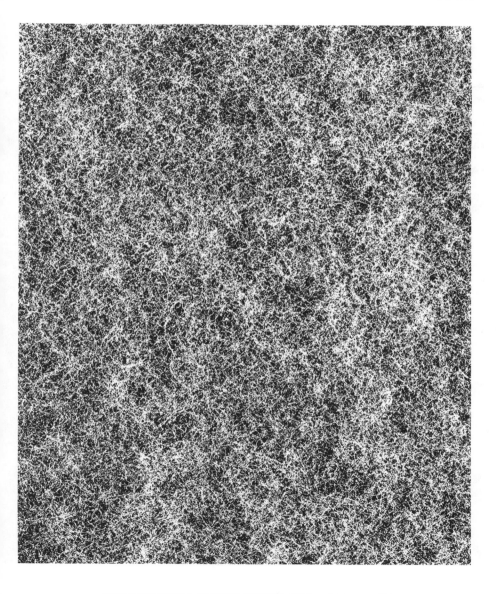

PROFESSIONAL PUBLICATIONS, INC. ● Belmont, CA

MULTIPLY	BY	TO OBTAIN
cable lengths	120.	fathoms
	720.	feet
	219.456	meters
	0.1184	miles (nautical)
	240.	yards
calibers	0.0254	centimeters
	0.0008333	feet
	0.01	inches
calories	1.	gram-cal
	1.0×10^{-7}	joules
	0.001	kilogram-cal
	0.001	large calories
	1.	small calories
calories (15°C)	4.1858	joules
calories (20°C)	4.1819	joules
calories (IST)	0.001	kilocalories (IST)
calories (IST)	0.0039683	BTU (IST)
	0.0039710	BTU (thermo)
	1.	calories (IST)
	0.999232	calories (mean)
	1.00067	calories (thermo)
	2.6132×10^{19}	electronvolts
	3.088	foot-lbf
	1.559×10^{-6}	horsepower-hr (electric)
	4.1868	joules
	0.001163	watt-hr
calories (large)	1.	kilogram-cal
calories (mean)	0.0039683	BTU (mean)
	1.00077	calories (IST)
	1.00144	calories (thermo)
	0.0014603	cubic foot-atm
	3.0904	foot-lbf
	4.19002	joules
	0.42726	kilogram·force-meter
	0.002206	pound-cal
	0.0011639	watt-hr

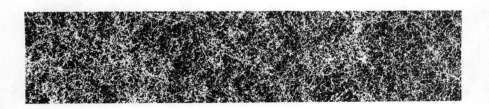

PROFESSIONAL PUBLICATIONS, INC. ● Belmont, CA

MULTIPLY	BY	TO OBTAIN
calories (thermo)	0.0039657	BTU (IST)
	0.0039683	BTU (thermo)
	0.999331	calories (IST)
	0.998563	calories (mean)
	0.02143	cubic foot-lbf/in^2
	2.611×10^{19}	electron volts
	3.086	foot-lbf
	1.559×10^{-6}	horsepower-hr (electric)
	4.184	joules
	0.04129	liter-atm
	0.0011622	watt-hr
calories/cm-sec-°C	241.9	BTU-ft/ft^2-hr-°F
calories/cm^2-min	697.33	watts/m^2
calories/cm^2-sec-°C	7373	BTU/ft^2-hr-°F
calories/g	1.8	BTU/lbm
	1.	CHU/lbm
calories/g-°C	1.	BTU/lbm-°F
calories/sec	4.184	watts
candlepower (see candles)		
candles (int'l)	1.	bougie decimales
	0.104167	carcel units
	1.	English sperm candles
	1.11111	hefner units
	1.	lumens/steradian
	1.	pentane candles
candles/cm^2	6.45156	candles/in^2
	3.14159	lamberts
	1.	lumens/cm^2-sterad
	929.026	lumens/ft^2-sterad
	3141.59	millilamberts
	1.	stilbs
candles/ft^2	33.8158	apostilbs
	0.0069444	candles/in^2
	10.764	candles/m^2
	3.14155	foot-lamberts
	0.0033816	lamberts
	0.0010764	lumens/cm^2-sterad
	1.	lumens/ft^2-sterad
	0.0010764	stilbs

PROFESSIONAL PUBLICATIONS, INC. • Belmont, CA

MULTIPLY	BY	TO OBTAIN
candles/in^2	4869.5	apostilbs
	0.155	candles/cm^2
	144.	candles/ft^2
	1550.	candles/m^2
	452.38	foot-lamberts
	0.48695	lamberts
	0.155	lumens/cm^2-sterad
	144.	lumens/ft^2-sterad
	486.95	millilamberts
	0.155	stilbs
candles/m^2	3.14159	apostilbs
	0.092903	candles/ft^2
	6.4516×10^{-4}	candles/in^2
	0.29186	foot-lamberts
	3.14159×10^{-4}	lamberts
	1.0×10^{-4}	lumens/cm^2-sterad
	0.092903	lumens/ft^2-sterad
	1.	nits
	1.0×10^{-4}	stilbs
carats (metric)	0.11288	drams (avoir)
	0.051441	drams (troy)
	3.0865	grains
	0.2	grams
	200.	milligrams
	0.0070548	ounces (avoir)
	0.0064302	ounces (troy)
	0.12860	pennyweights
	4.4093×10^{-4}	pounds (avoir)
	5.3585×10^{-4}	pounds (troy)
	0.15432	scruples
	0.0068571	tons (assay)
carats (metric)	0.97468	carats (troy)
carats (troy)	1.0260	carats (metric)
	0.11581	drams (avoir)
	0.052778	drams (troy)
	3.16667	grains
	205.197	milligrams
	0.007238	ounces (avoir)
	0.006597	ounces (troy)
carcels	9.6	bougie decimales
	9.6	candles (international)
	9.6	English sperm candles
	10.667	hefner units
	9.6	lumens/sterad
	9.6	pentane candles

PROFESSIONAL PUBLICATIONS, INC. ● Belmont, CA

MULTIPLY	BY	TO OBTAIN
celos	30.48	centimeters/sec^2
	1.	feet/sec^2
	0.031081	g (gravity)
	1.0973	kilometers/hr-sec
	0.304801	meters/sec^2
	40.909	miles/hr-min
	0.6818	miles/hr-sec
centals	0.89286	hundredweights (long)
	1.	hundredweights (short)
	45.3592	kilograms
	1600.	ounces (avoir)
	100.	pounds (avoir)
	0.178571	quarters (long)
	0.2	quarters (short)
	0.453592	quintals
	3.1081	slugs
	0.044643	tons (long)
	0.045359	tons (metric)
	0.05	tons (short)
centares	2.47105×10^{-4}	acres
	0.01	ares
	1.0×10^{-4}	hectares
	10,000.	square centimeters
	0.00247105	square chains (Gunter's or surveyor's)
	10.764	square feet
	1550.	square inches
	1.0×10^{-6}	square kilometers
	24.71	square links (Gunter's or surveyor's)
	1.	square meters
	3.861×10^{-7}	square miles
	0.0395367	square rods (also perches or poles)
	1.196	square yards
centigrams	0.0056438	drams (avoir)
	0.0025721	drams (troy)
	0.15432	grains
	0.01	grams
	3.5274×10^{-4}	ounces (avoir)
	3.2151×10^{-4}	ounces (troy)
	0.0064302	pennyweights
	2.2046×10^{-5}	pounds (avoir)
	2.6792×10^{-5}	pounds (troy)
	0.0077162	scruples

MULTIPLY	BY	TO OBTAIN
centimeters	1.0×10^8	angstroms
	1.1811	barleycorns
	39.37	calibers
	0.021872	cubits
	0.001	decameters (also dekameters)
	0.1	decimeters
	0.52493	digits
	56.693	douziemes
	2.3611	ems (pica)
	0.0054681	fathoms
	0.0328084	feet
	0.098425	hands
	0.393701	inches
	1.0×10^{-5}	kilometers
	4.7244	lines
	0.04971	links (Gunter's or surveyor's)
	0.03281	links (Ramsden's or engineer's)
	0.01	meters
	1.0×10^4	micrometers
	1.0×10^{10}	micromicrons
	10,000.	microns
	6.214×10^{-6}	miles
	10.	millimeters
	1.0×10^7	millimicrons
	393.701	mils
	0.175	nails
	1.0×10^7	nanometers
	0.0131	paces
	0.131	palms
	0.0019884	perches
	28.452	points
	0.0019884	rods
	0.04374	spans
	15,531.6	wavelengths red line cadmium
	0.010936	yards
centimeters Hg at 0°C	0.013158	atmospheres
	0.013332	bars
	13,332.	dynes/cm^2
	13.595	grams/cm^2
	0.3937	inches Hg at 32°F
	0.013595	kilograms/cm^2
	135.95	kilograms/m^2
	1.3595×10^{-4}	kilograms/mm^2
	3.09379	ounces/in^2
	27.845	pounds/ft^2
	0.19337	pounds/in^2
	0.013923	tons (short)/ft^2
	9.6684×10^{-5}	tons (short)/in^2

MULTIPLY	BY	TO OBTAIN
centimeters H_2O at 4°C	9.6781×10^{-4}	atmospheres
	9.8064×10^{-4}	bars
	980.64	dynes/cm^2
	1.0	grams/cm^2
	1.0	kilograms/m^2
	2.0481	pounds/ft^2
	0.014223	pounds/in^2
centimeters/sec	118.11	feet/hr
	1.9685	feet/min
	0.032808	feet/sec
	0.036	kilometers/hr
	6.0×10^{-4}	kilometers/min
	0.019425	knots
	0.6	meters/min
	0.01	meters/sec
	0.02237	miles/hr
	3.727×10^{-4}	miles/min
centimeters/sec^2	0.03281	feet/sec^2
	0.00101972	g (gravity)
	1.	galileos
	0.3937	inches/sec^2
	0.036	kilometers/hr-sec
	0.01	meters/sec^2
	1.34216	miles/hr-min
	0.02237	miles/hr-sec
centipoises	0.01	dyne-sec/cm^2
	0.01	gram/sec-cm
	1.01972×10^{-4}	kilogram (force)-sec/m^2
	3.6	kilograms/m-hr
	0.001	pascal-second
	0.01	poises
	2.4191	pounds (mass)/ft-hr
	6.7197×10^{-4}	pounds (mass)/ft-sec
	2.089×10^{-5}	pound-sec/ft^2
	1.450×10^{-7}	pound-sec/in^2
	2.4193	poundal-hr/ft^2
	6.7197×10^{-4}	poundal-sec/ft^2
centistokes	0.01	square centimeters/sec
	0.03875	square feet/hr
	1.0764×10^{-5}	square feet/sec
	0.001550	square inches/sec
	0.0036	square meters/hr
	1.0×10^{-6}	square meters/sec
	0.01	stokes

MULTIPLY	BY	TO OBTAIN
centuries	10.	decades
	0.1	millennia
	100.	years
chains	0.66	chains (Ramsden's or engineer's)
(Gunter's or surveyor's)	66.	feet
	0.1	furlongs
	792.	inches
	0.0201168	kilometers
	100.	links (Gunter's or surveyor's)
	66.	links (Ramsden's or engineer's)
	20.1168	meters
	0.0125	miles (statute)
	26.4	paces
	4.	rods (also perches or poles)
	22.	yards
chains	1.51515	chains (Gunter's or surveyor's)
(Ramsden's or engineer's)	100.	feet
	0.151515	furlongs
	1200.	inches
	0.03048	kilometers
	151.515	links (Gunter's or surveyor's)
	100.	links (Ramsden's or engineer's)
	30.48	meters
	0.01894	miles (statute)
	40.	paces
	6.0606	rods (also perches or poles)
	33.3333	yards
chaldrons	36.	bushels
	1.2686×10^6	cubic centimeters
	44.800	cubic feet
	77,415.	cubic inches
	1.2686	cubic meters
	1.6593	cubic yards
	1.2686	steres
CHU	1.8	BTU
	454.	gram-cal
	1897.8	joules
circles	360.	degrees
	400.	grades
	21,600.	minutes
	4.	quadrants
	6.28319	radians
	1.296×10^6	seconds
	12.	signs

MULTIPLY	BY	TO OBTAIN
circular centimeters	0.0010764	circular feet
	0.1550	circular inches
	0.785398	square centimeters
	0.121736	square inches
	7.854×10^{-5}	square meters
circular feet	929.03	circular centimeters
	144.	circular inches
	729.66	square centimeters
	0.785398	square feet
	113.1	square inches
	0.072966	square meters
circular inches	6.452	circular centimeters
	0.0069444	circular feet
	645.161	circular millimeters
	1.0×10^{5}	circular mils
	5.06707	square centimeters
	0.00545415	square feet
	0.785398	square inches
	5.06707×10^{-4}	square meters
	506.707	square millimeters
	785,403.	square mils
	6.0606×10^{-4}	square yards
circular millimeters	0.00155	circular inches
	15,500.	circular mils
	0.00785398	square centimeters
	8.45395×10^{-6}	square feet
	0.00121737	square inches
	7.85398×10^{-6}	square meters
	7.85398	square millimeters
	1217.38	square mils
	9.3933×10^{-6}	square yards
circular mils	1.0×10^{-6}	circular inches
	6.4516×10^{-5}	circular millimeters
	5.06707×10^{-6}	square centimeters
	5.45415×10^{-9}	square feet
	7.85398×10^{-7}	square inches
	5.06707×10^{-10}	square meters
	5.06707×10^{-4}	square millimeters
	0.785398	square mils
cords	1536.	board-ft
	8.	cord-ft
	3.6246×10^{6}	cubic centimeters
	128.	cubic feet
	221,184.	cubic inches
	3.625	cubic meters
	4.7407	cubic yards
	3.6246	steres

PROFESSIONAL PUBLICATIONS, INC. ● Belmont, CA

MULTIPLY	BY	TO OBTAIN
cord-ft	192.	board-ft
	0.125	cords
	453,069.	cubic centimeters
	16.	cubic feet
	27,648.	cubic inches
	0.45307	cubic meters
	0.59260	cubic yards
	0.45307	steres
coulombs	0.1	abcoulombs
	2.7778×10^{-4}	ampere-hr
	6.281×10^{18}	electronic charges
	836,680.	electrostatic ft-lbf-sec
	2.998×10^9	electrostatic units
	1.0363×10^{-5}	faradays
	2.998×10^9	statcoulombs
coulombs/cm^2	0.1	abcoulombs/cm^2
	6.452	coulombs/in^2
	10,000.	coulombs/m^2
	2.998×10^9	statcoulombs/cm^2
coulombs/in^2	0.0155	abcoulombs/cm^2
	0.155	coulombs/cm^2
	1550.	coulombs/m^2
	4.647×10^8	statcoulombs/cm^2
coulombs/m^2	1.0×10^{-5}	abcoulombs/cm^2
	1.0×10^{-4}	coulombs/cm^2
	6.452×10^{-4}	coulombs/in^2
	2.998×10^5	statcoulombs/cm^2
coulombs/sec	1.	amperes
coulombs/volt	1.	farads
crinals	10,000.	dynes
	0.0101972	kilograms (force)
	0.1	newtons
	0.022481	pounds
	0.72330	poundals
	1.0×10^{-4}	sthènes

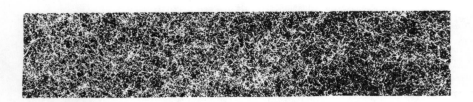

MULTIPLY	BY	TO OBTAIN
cubic centimeters	4.2378×10^{-4}	board-ft
	0.1	centiliters
	3.5315×10^{-5}	cubic feet
	0.061024	cubic inches
	1.0×10^{-6}	cubic meters
	1000.	cubic millimeters
	0.27051	drams
	2.3×10^{-4}	gallons (U.S., dry)
	2.6417×10^{-4}	gallons (U.S., liquid)
	0.0084535	gills
	15.432	grains
	0.001	liters
	1000.	microliters
	1.	milliliters
	16.231	minims
	0.033814	ounces (U.S., liquid)
	0.0018162	pints (U.S., dry)
	0.0021134	pints (U.S., liquid)
	9.0808×10^{-4}	quarts (U.S., dry)
	0.0010567	quarts (U.S., liquid)
	1.0×10^{-6}	steres
cubic centimeter-atm	9.604×10^{-5}	BTU
	3.531×10^{-5}	cubic foot-atm
	1.01325×10^{6}	dyne-cm
	1.01325×10^{6}	ergs
	0.07473	foot-lbf
	2.4045	foot-poundals
	0.02421	gram-cal
	1033.	gram-cm
	3.774×10^{-8}	horsepower-hr (U.S.)
	0.101325	joules
	2.421×10^{-5}	kilogram-cal
	0.010332	kilogram-m
	2.815×10^{-8}	kilowatt-hr
	0.001	liter-atm
	1.01325	megalergs
	2.815×10^{-5}	watt-hr
cubic centimeters/sec	0.00211887	cubic feet/min
	3.53148×10^{-5}	cubic feet/sec
	7.8476×10^{-5}	cubic yards/min
	0.0158502	gallons (U.S., liquid)/min
	2.64185×10^{-4}	gallons (U.S., liquid)/sec
	0.06	liters/min
	0.001	liters/sec
	0.0017657	miner's inches*
	0.13228	pounds H_2O/min

* ID, KS, ND, NE, NM, NV, SD, UT

PROFESSIONAL PUBLICATIONS, INC. ● Belmont, CA

MULTIPLY	BY	TO OBTAIN
cubic feet	2.2957×10^{-5}	acre-ft
	0.23748	barrels (31.5 U.S. gallons)
	0.1781	barrels (42 U.S. gallons)
	12.	board-ft
	0.80356	bushels
	0.022321	chaldrons
	0.0625	cord-ft
	0.0078125	cords
	28,317.	cubic centimeters
	1728.	cubic inches
	0.028317	cubic meters
	0.037037	cubic yards
	7,660.	drams
	0.83117	firkins
	6.4285	gallons (U.S., dry)
	7.481	gallons (U.S., liquid)
	239.38	gills
	0.11874	hogsheads
	0.028317	kiloliters
	28.3169	liters
	957.51	ounces (U.S., liquid)
	3.2143	pecks
	51.428	pints (U.S., dry)
	59.844	pints (U.S., liquid)
	25.714	quarts (U.S., dry)
	29.922	quarts (U.S., liquid)
	0.028317	steres
cubic feet/day	6.944×10^{-4}	cubic feet/min
	1.157×10^{-5}	cubic feet/sec
	1.966×10^{-5}	cubic meters/min
	0.31169	gallons (U.S., liquid)/hr
	3.277×10^{-4}	liters/sec
cubic feet/hr	0.016667	cubic feet/min
	2.7778×10^{-4}	cubic feet/sec
	4.719×10^{-4}	cubic meters/min
	7.481	gallons (U.S., liquid)/hr
	0.007866	liters/sec

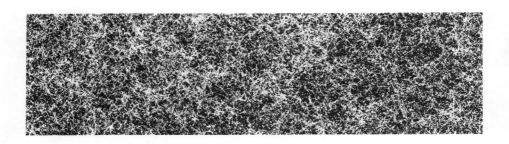

PROFESSIONAL PUBLICATIONS, INC. • Belmont, CA

MULTIPLY	BY	TO OBTAIN
cubic feet/mile^2	4.305×10^{-7}	inches of runoff
cubic feet/min	256.47	barrels (42 U.S. gallons)/day
	10.687	barrels (42 U.S. gallons)/hr
	0.1781	barrels (42 U.S. gallons)/min
	471.95	cubic centimeters/sec
	1440.	cubic feet/day
	60.	cubic feet/hr
	0.0166667	cubic feet/sec
	28.800	cubic inches/sec
	0.02832	cubic meters/min
	0.0370373	cubic yards/min
	10,772.	gallons (U.S., liquid)/day
	448.83	gallons (U.S., liquid)/hr
	7.48052	gallons (U.S., liquid)/min
	0.124675	gallons (U.S., liquid)/sec
	28.3168	liters/min
	0.47195	liters/sec
	0.833333	miner's inches*
	3741.	pounds H_2O/hr
	62.43	pounds H_2O/min
cubic feet/sec	1.9835	acre-ft/day
	0.99173	acre-inches/hr
	15,389.	barrels (42 U.S. gallons)/day
	641.19	barrels (42 U.S. gallons)/hr
	10.686	barrels (42 U.S gallons)/min
	28,317.	cubic centimeters/sec
	86,400.	cubic feet/day
	3600.	cubic feet/hr
	60.	cubic feet/min
	1728.	cubic inches/sec
	1.6990	cubic meters/min
	2.22224	cubic yards/min
	646,317.	gallons (U.S., liquid)/day
	26,930.	gallons (U.S., liquid)/hr
	448.83	gallons (U.S., liquid)/min
	7.4805	gallons (U.S., liquid)/sec
	1699.	liters/min
	28.317	liters/sec
	0.64632	MGD (millions of gallons/day)
	50.	miner's inches*
	224,460.	pounds H_2O/hr
	3745.8	pounds H_2O/min

* ID, KS, ND, NE, NM, NV, SD, UT

PROFESSIONAL PUBLICATIONS, INC. ● Belmont, CA

MULTIPLY	BY	TO OBTAIN
cubic foot-atm	2.720	BTU
	28,314.	cubic centimeter-atm
	2.8692×10^{10}	dyne-cm
	2.8692×10^{10}	ergs
	2116.2	foot-lbf
	68,087.	foot-poundals
	685.4	gram-cal
	2.926×10^{7}	gram-cm
	0.001069	horsepower-hr (U.S.)
	2869.2	joules
	0.6854	kilogram-cal
	292.58	kilogram-m
	7.970×10^{-4}	kilowatt-hr
	28.317	liter-atm
	28,694.	megalergs
	0.7970	watt-hr
cubic inches	1.3743×10^{-4}	barrels (31.5 U.S. gallons)
	0.006944	board-ft
	4.6503×10^{-4}	bushels
	1.63871	centiliters
	16.3871	cubic centimeters
	5.7870×10^{-4}	cubic feet
	1.6387×10^{-5}	cubic meters
	16,387.1	cubic millimeters
	2.1434×10^{-5}	cubic yards
	4.4329	drams
	4.810×10^{-4}	firkins
	0.003720	gallons (U.S., dry)
	0.004329	gallons (U.S., liquid)
	0.13853	gills
	0.0163871	liters
	16.3871	milliliters
	265.97	minims
	0.5541	ounces (U.S., liquid)
	0.0018601	pecks
	0.029762	pints (U.S., dry)
	0.034632	pints (U.S., liquid)
	0.014881	quarts (U.S., dry)
	0.017316	quarts (U.S., liquid)
	1.6387×10^{-5}	steres
cubic inches/sec	0.0347	cubic feet/min
	9.832×10^{-4}	cubic meters/min
	0.25974	gallons (U.S., liquid)/min
	0.016387	liters/sec

MULTIPLY	BY	TO OBTAIN
cubic meters	8.3864	barrels (31.5 U.S. gallons)
	6.2898	barrels (42 U.S. gallons)
	423.78	board-ft
	28.378	bushels
	100,000.	centiliters
	0.78827	chaldrons
	2.2072	cord-ft
	0.2759	cords
	1.0×10^6	cubic centimeters
	35.3147	cubic feet
	61,024.	cubic inches
	1.3080	cubic yards
	29.353	firkins
	227.0	gallons (U.S., dry)
	264.172	gallons (U.S., liquid)
	8453.5	gills
	4.1932	hogsheads
	1.	kiloliters
	1000.	liters
	100,000.	milliliters
	33,814.	ounces (U.S., liquid)
	113.51	pecks
	1816.2	pints (U.S., dry)
	2113.4	pints (U.S., liquid)
	908.08	quarts (U.S., dry)
	1056.7	quarts (U.S., liquid)
	1.	steres
cubic meters/min	35.314	cubic feet/min
	0.5886	cubic feet/sec
	264.17	gallons (U.S., liquid)/min
	16.667	liters/sec
cubic meters/m-day	80.52	gallons/ft-day
cubic meters/m²-day	24.54	gallons/ft²-day
cubic millimeters	0.001	cubic centimeters
	6.1024×10^{-5}	cubic inches
	2.7051×10^{-4}	drams
	1.	microliters
	0.001	milliliters
	0.01623	minims
	3.3814×10^{-5}	ounces

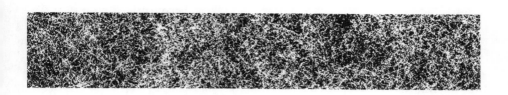

PROFESSIONAL PUBLICATIONS, INC. • Belmont, CA

MULTIPLY	BY	TO OBTAIN
cubic yards	6.2×10^{-4}	acre-ft
	6.4119	barrels (31.5 U.S. gallons)
	4.8089	barrels (42 U.S. gallons)
	324.	board-ft
	21.696	bushels
	0.60267	chaldrons
	1.6875	cord-ft
	0.21094	cords
	764,555.	cubic centimeters
	27.	cubic feet
	46,656.	cubic inches
	0.76456	cubic meters
	22.442	firkins
	173.57	gallons (U.S., dry)
	202.	gallons (U.S., liquid)
	6463.2	gills
	3.2059	hogsheads
	0.76456	kiloliters
	764.56	liters
	25,853.	ounces (U.S., liquid)
	86.785	pecks
	1388.6	pints (U.S., dry)
	1616.	pints (U.S., liquid)
	694.28	quarts (U.S., dry)
	807.9	quarts (U.S., liquid)
	0.76456	steres
cubic yards/min	12,743.	cubic centimeters/sec
	27.	cubic feet/min
	0.45	cubic feet/sec
	202.	gallons (U.S., liquid)/min
	3.366	gallons (U.S., liquid)/sec
	764.6	liters/min
	12.74	liters/sec
	22.5	miner's inches*
	1685.6	pounds H_2O/min
cubits	45.720	centimeters
	1.5	feet
	18.	inches
	0.45720	meters
	0.5	yards
cups	8.	ounces (U.S., liquid)
	0.5	pints (U.S., liquid)
curies	3.7×10^{10}	becquerels
	3.7×10^{10}	disintegrations/sec
cycles/sec	1.	hertz

* ID, KS, ND, NE, NM, NV, SD, UT

PROFESSIONAL PUBLICATIONS, INC. • Belmont, CA

MULTIPLY	BY	TO OBTAIN
daltons	1.66×10^{-24}	grams
	1.66×10^{-27}	kilograms
	5.8563×10^{-26}	ounces (avoir)
	3.6602×10^{-27}	pounds (avoir)
darcys	1.062×10^{-11}	square feet
	9.86923×10^{-13}	square meters
days (sidereal)	0.997270	days (solar)
	23.9345	hours (solar)
	1436.1	minutes (solar)
	0.033771	months (lunar)
	0.032787	months (mean calendar)
	86,400.	seconds (sidereal)
	86,164.	seconds (solar)
	0.14247	weeks
	0.0027248	years (leap)
	0.0027303	years (sidereal)
	0.0027304	years (solar)
days (solar)	1.00274	days (sidereal)
	0.071429	fortnights
	24.	hours (solar)
	1440.	minutes (solar)
	0.033863	months (lunar)
	0.032877	months (mean calendar)
	86,637.	seconds (sidereal)
	86,400.	seconds (solar)
	0.14286	weeks
	0.0027322	years (leap)
	0.0027378	years (sidereal)
	0.0027379	years (solar)
days/kg	0.45359	days/lbm
	0.03110	days/ounces (troy)
	0.024	hours/g
	10.886	hours/lbm
days/lbm	2.2046	days/kg
	0.068573	days/ounce (troy)
	0.052911	hours/g
	24.	hours/lbm
days/ounce (troy)	32.151	days/kg
	14.583	days/lbm
	0.77162	hours/g
	350.	hours/lbm
debye units	$1.0 \wedge 10^{18}$	electrostatic units

MULTIPLY	BY	TO OBTAIN
decades	0.1	centuries
	0.01	millennia
	10.	years
decagrams	1.	dekagrams
	10.	grams
	0.35274	ounces (avoir)
	0.022046	pounds (avoir)
decaliters	0.28378	bushels
	610.24	cubic inches
	1.	dekaliters
	2.6417	gallons (U.S., liquid)
	10.	liters
decameters	1000.	centimeters
	1.	dekameters
	32.8084	feet
	393.70	inches
	10.	meters
	10.936	yards
decasteres	10.	cubic meters
	1.	dekasteres
	10.	steres
decibels	0.1	bels
decigrams	0.1	grams
	0.0035274	ounces (avoir)
	2.2046×10^{-4}	pounds (avoir)
deciliters	6.1024	cubic inches
	0.1	liters
decimeters	10.	centimeters
	0.328083	feet
	3.9370	inches
	0.1	meters
	0.10936	yards
decisteres	0.1	cubic meters
	0.1	steres
degrees (angular)	1.11111	grades
	17.7778	mils
	60.	minutes
	0.0111111	quadrants
	0.017453	radians
	0.00277778	revolutions
	3600.	seconds
	0.033333	signs

MULTIPLY	BY	TO OBTAIN
degrees, Clarke	1.	grains/Imperial gallon
	0.8333	grains/gallon (U.S.)
degrees/cm	30.480	degrees/ft
	2.54	degrees/in
	60.	minutes/cm
	0.017453	radians/cm
degrees/ft	0.032808	degrees/cm
	0.083333	degrees/in
	1.9686	minutes/cm
	5.726×10^{-4}	radians/cm
degrees/in	0.3937	degrees/cm
	12.	degrees/ft
	23.622	minutes/cm
	0.0068714	radians/cm
degrees/sec	0.017453	radians/sec
	240.	revolutions/day
	0.16667	revolutions/min
	0.0027778	revolutions/sec
dekagrams	1.	decagrams
dekameters	1.	decameters
digits	1.905	centimeters
	0.0625	feet
	0.33333	hands
	0.75	inches
	0.25	palms
	0.083333	spans
disintegrations/sec	1.	becquerels
	2.703×10^{-11}	curies
doors	1.0×10^{26}	square feet
douziemes	0.020833	barleycorns
	0.017639	centimeters
	5.7870×10^{-4}	feet
	0.0069444	inches
	0.083333	lines
dozens	0.08333	gross
	12.	units

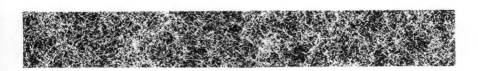

PROFESSIONAL PUBLICATIONS, INC. ● Belmont, CA

MULTIPLY	BY	TO OBTAIN
drams (avoir)	8.8592	carats
	177.1845	centigrams
	0.45573	drams (troy)
	27.344	grains
	1.771845	grams
	0.0625	ounces (avoir)
	0.056966	ounces (troy)
	1.1393	pennyweights
	0.0039063	pounds (avoir)
	0.0047472	pounds (troy)
	1.3672	scruples
	0.06075	tons (assay)
drams (troy)	19.440	carats
	388.79	centigrams
	2.1943	drams (avoir)
	60.	grains
	3.8879	grams
	0.13714	ounces (avoir)
	0.125	ounces (troy)
	2.5	pennyweights
	0.0085714	pounds (avoir)
	0.0104167	pounds (troy)
	3.	scruples
	0.1333	tons (assay)
drams (U.S., liquid)	3.6967	cubic centimeters
	1.3055×10^{-4}	cubic feet
	0.225586	cubic inches
	9.7656×10^{-4}	gallons (U.S., liquid)
	0.03125	gills
	0.0036967	liters
	3.6967	milliliters
	60.	minims
	0.125	ounces (U.S., liquid)
	0.0078125	pints (U.S., liquid)
	0.00390625	quarts (U.S., liquid)
drams/ounce (U.S., liquid)	13.672	grains/ounce (U.S., liquid)
	59.913	kilograms/m^3
	8.	ounces/gallon (U.S., liquid)
	0.034632	ounces (U.S., liquid)/in^3
	2.	ounces/quart (U.S., liquid)
	3.7403	pounds/ft^3
	0.0021645	pounds/in^3

MULTIPLY	BY	TO OBTAIN
dynes	1.0×10^{-4}	crinals
	0.015737	grains
	0.0010197	grams (force)
	1.0×10^{-5}	joules/m
	1.0197×10^{-6}	kilograms (force)
	1.0×10^{-5}	newtons
	2.2481×10^{-6}	pounds
	7.233×10^{-5}	poundals
	1.0×10^{-8}	sthènes
	1.0036×10^{-9}	tons (long)
	1.12405×10^{-9}	tons (short)
dynes/cm	1.	ergs/cm^2
	0.01	ergs/mm^2
	2.5901	milligram wt/in
	0.10197	milligram wt/mm
dynes/cm^2	9.8692×10^{-7}	atmospheres
	1.0×10^{-6}	bars
	7.500×10^{-5}	centimeters Hg at 0°C
	0.0010197	grams/cm^2
	2.953×10^{-5}	inches Hg at 32°F
	1.0197×10^{-6}	kilograms/cm^2
	0.010197	kilograms/m^2
	1.0197×10^{-8}	kilograms/mm^2
	2.3206×10^{-4}	ounces/in^2
	0.1	pascals
	0.0020885	pounds/ft^2
	1.4504×10^{-5}	pounds/in^2
	1.0443×10^{-6}	tons (short)/ft^2
	7.2519×10^{-9}	tons (short)/in^2
dynes/cm^3	0.00101972	grams (force)/cm^3
	5.894×10^{-4}	ounces/in^3
	3.684×10^{-5}	pounds/in^3
	0.0011853	poundals/in^3

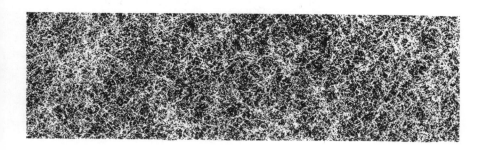

PROFESSIONAL PUBLICATIONS, INC. • Belmont, CA

MULTIPLY	BY	TO OBTAIN
dyne-cm	9.478×10^{-11}	BTU
	9.869×10^{-7}	cubic centimeter-atm
	3.485×10^{-11}	cubic foot-atm
	1.	ergs
	7.3756×10^{-8}	foot-lbf
	2.3730×10^{-6}	foot-poundals
	2.389×10^{-8}	gram-cal
	0.0010197	gram-cm
	3.725×10^{-14}	horsepower-hr (U.S.)
	8.8511×10^{-7}	inch-lbf
	1.0×10^{-7}	joules
	2.3889×10^{-11}	kilogram-cal
	1.01972×10^{-8}	kilogram-m
	2.7778×10^{-14}	kilowatt-hr
	9.869×10^{-10}	liter-atm
	1.0×10^{-6}	megalergs
	2.7778×10^{-11}	watt-hr
dyne-cm/sec	5.683×10^{-9}	BTU/min
	9.4709×10^{-11}	BTU/sec
	1.	ergs/sec
	2.655×10^{-4}	foot-lbf/hr
	4.4254×10^{-6}	foot-lbf/min
	7.3756×10^{-8}	foot-lbf/sec
	0.001020	gram-cm/sec
	1.3596×10^{-10}	horsepower (metric)
	1.3412×10^{-10}	horsepower (U.S.)
	1.0×10^{-7}	joules/sec
	1.4333×10^{-9}	kilogram-cal/min
	2.3888×10^{-11}	kilogram-cal/sec
	1.020×10^{-8}	kilogram-m/sec
	1.0×10^{-10}	kilowatts
	6.6845×10^{-5}	lumens
	1.0×10^{-7}	watts
dyne-sec/cm^2	100.	centipoises
	1.	grams/cm-sec
	0.010197	kilograms (force)-sec/m^2
	360.	kilograms/m-hr
	1.	poises
	241.93	pounds (mass)/ft-hr
	0.067204	pounds (mass)/ft-sec
	0.002089	pound-sec/ft^2
	1.450×10^{-5}	pound-sec/in^2
	241.93	poundal-hr/ft^2
	0.067204	poundal-sec/ft^2

MULTIPLY	BY	TO OBTAIN
edisons	100.	amperes
electronic charges	1.602×10^{-20}	abcoulombs
	4.4506×10^{-23}	ampere-hr
	1.6022×10^{-19}	coulombs
	1.3321×10^{-13}	electrostatic ft-lbf-sec
	4.774×10^{-10}	electrostatic units
	1.660×10^{-24}	faradays
	4.803×10^{-10}	statcoulombs
electron volts	1.074×10^{-9}	amu (nuclear)
	1.0×10^{-9}	BeV
	1.517×10^{-22}	BTU
	3.8293×10^{-20}	calories (thermo)
	1.6022×10^{-12}	ergs
	1.1817×10^{-19}	foot-lbf
	1.0×10^{-12}	GeV
	1.6022×10^{-19}	joules
	1.783×10^{-36}	kilograms (nuclear)
	4.45×10^{-26}	kilowatt-hr
	1.0×10^{-6}	MeV
electrostatic ft-lbf-sec	1.1952×10^{-7}	abcoulombs
	3.320×10^{-10}	ampere-hr
	1.1952×10^{-6}	coulombs
	7.5071×10^{12}	electronic charges
	1.2386×10^{-11}	faradays
	3583.2	statcoulombs
electrostatic units (esu)	3.336×10^{-6}	coulombs
	1.0	statcoulombs
ells	0.03125	bolts
	3.75	feet
	45.	inches
	1.143	meters
	20.0	nails (cloth)
	5.	quarters (cloth)
	0.0104167	skeins
	1.25	yards
ems (pica)	1.	bougie decimales
	0.42175	centimeters
	0.013837	feet
	0.166044	inches
	4.2175	millimeters
	166.044	mils
	12.	points

MULTIPLY	BY	TO OBTAIN
English sperm candles	1.	candles (int'l)
	0.104167	carcel units
	1.11111	hefner units
	1.	lumens/sterad
	1.	pentane candles
ergs	9.478×10^{-11}	BTU
	2.39×10^{-8}	calories
	9.869×10^{-7}	cubic centimeter-atm
	3.485×10^{-11}	cubic foot-atm
	1.	dyne-cm
	6.2415×10^{11}	electron volts
	7.376×10^{-8}	foot-lbf
	2.373×10^{-6}	foot-poundals
	2.389×10^{-8}	gram-cal
	0.0010197	gram-cm
	1.	$\text{gram-cm}^2/\text{sec}^2$
	3.725×10^{-14}	horsepower-hr (U.S.)
	1.0×10^{-7}	joules
	1.113×10^{-24}	kilograms
	2.3889×10^{-11}	kilogram-cal
	1.01972×10^{-8}	kilogram-m
	2.778×10^{-14}	kilowatt-hr
	9.869×10^{-10}	liter-atm
	1.0×10^{-6}	megalergs
	6.2415×10^{6}	MeV
	2.7778×10^{-11}	watt-m
ergs/cm^2	1.	dynes/cm
	0.01	ergs/mm^2
	2.5901	milligram wt/in
	1.0×10^{-3}	joules/m^2
	0.10197	milligram wt/mm
ergs/mm^2	100.	dynes/cm
	100.	ergs/cm^2
	259.01	milligram wt/in
	10.197	milligram wt/mm

PROFESSIONAL PUBLICATIONS, INC. ● Belmont, CA

MULTIPLY	BY	TO OBTAIN
ergs/sec	5.691×10^{-9}	BTU/min
	9.4845×10^{-11}	BTU/sec
	1.	dyne-cm/sec
	2.6549×10^{-4}	foot-lbf/hr
	4.4254×10^{-6}	foot-lbf/min
	7.3756×10^{-8}	foot-lbf/sec
	0.001020	gram-cm/sec
	1.3596×10^{-10}	horsepower (metric)
	1.3412×10^{-10}	horsepower (U.S.)
	1.0×10^{-7}	joules/sec
	1.4333×10^{-9}	kilogram-cal/min
	2.3888×10^{-11}	kilogram-cal/sec
	1.020×10^{-8}	kilogram-m/sec
	1.0×10^{-10}	kilowatts
	6.6845×10^{-5}	lumens
	1.0×10^{-7}	watts
erg-cm	2.7778×10^{-14}	kilowatt/hr

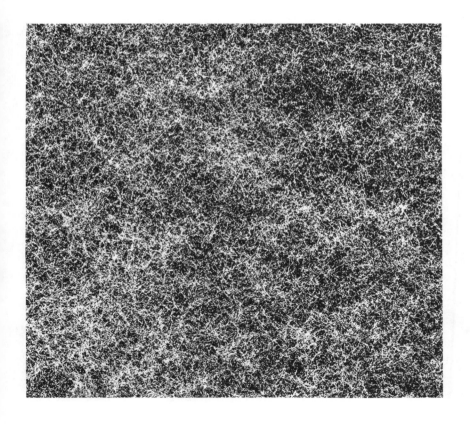

PROFESSIONAL PUBLICATIONS, INC. ● Belmont, CA

MULTIPLY	BY	TO OBTAIN
faradays	9648.7	abcoulombs
	26.802	ampere-hr
	96,487.	coulombs
	6.0226×10^{23}	electronic charges
	8.0746×10^{10}	electrostatic ft-lbf-sec
	2.8926×10^{14}	statcoulombs
faradays/sec	9650.	abamperes
	96,500.	amperes
	9.65×10^{7}	milliamperes
	2.89303×10^{14}	statamperes
farads	1.0×10^{-9}	abfarads
	1.	coulombs/volt
	1.0×10^{6}	microfarads
	1.0×10^{12}	micromicrofarads
	1.0×10^{12}	picofarads
	8.98755×10^{11}	statfarads
fathoms	0.0083333	cable lengths
	182.88	centimeters
	6.	feet
	18.	hands
	72.	inches
	3.289×10^{-4}	leagues (nautical)
	1.8288	meters
	9.8684×10^{-4}	miles (nautical)
	0.363636	rods (also perches or poles)
	2.	yards

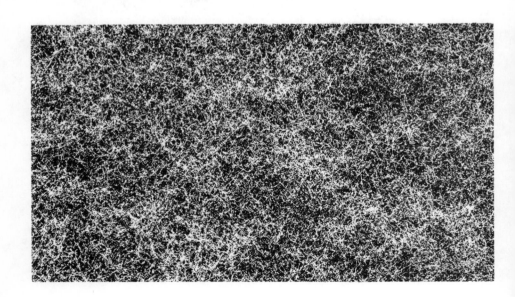

PROFESSIONAL PUBLICATIONS, INC. • Belmont, CA

MULTIPLY	BY	TO OBTAIN
feet	3.0480×10^9	angstroms
	2.0375×10^{-12}	astronomical units
	36.	barleycorns
	0.0083333	bolts
	1.3889×10^{-3}	cable lengths
	1200.	calibers
	30.48	centimeters
	0.0151515	chains (Gunter's or surveyor's)
	0.01	chains (Ramsden's or engineer's)
	0.666667	cubits
	0.03048	decameters
	3.048	decimeters
	16.	digits
	1728.	douziemes
	0.26667	ells
	72.27	ems (pica)
	0.166667	fathoms
	3.048×10^{14}	fermis
	0.00151515	furlongs
	3.	hands
	12.	inches
	3.048×10^{-4}	kilometers
	5.4825×10^{-5}	leagues (nautical)
	6.31313×10^{-5}	leagues (statute)
	144.	lines
	1.51515	links (Gunter's or surveyor's)
	1.	links (Ramsden's or engineer's)
	0.3048	meters
	3.048×10^5	micrometers
	304,800.	microns
	12,000.	mils
	1.6447×10^{-4}	miles (nautical)
	1.8939×10^{-4}	miles (statute)
	304.80	millimeters
	5.333	nails
	3.048×10^8	nanometers
	0.4	paces
	4.	palms
	0.060606	perches
	867.24	points
	7.57575×10^{-4}	rambles
	0.060606	rods
	0.05	ropes
	0.00277778	skeins
	1.3333	spans
	3.048×10^{11}	stigmas
	3.048×10^9	tenthmeters
	0.36	varas
	473,404.	wavelengths red line cadmium
	3.042×10^{12}	X-units
	0.333333	yards
	3.048×10^{14}	yukawas

PROFESSIONAL PUBLICATIONS, INC. • Belmont, CA

MULTIPLY	BY	TO OBTAIN
feet^2	(see square feet)	
feet^3	(see cubic feet)	
feet/hr	0.0084667	centimeters/sec
	0.016667	feet/min
	2.77778×10^{-4}	feet/sec
	3.0480×10^{-4}	kilometers/hr
	1.6447×10^{-4}	knots
	0.00508	meters/min
	8.4667×10^{-5}	meters/sec
	1.8939×10^{-5}	miles/hr
	3.16×10^{-6}	miles/min
feet/min	0.508	centimeters/sec
	60.	feet/hr
	0.016667	feet/sec
	0.01829	kilometers/hr
	3.048×10^{-4}	kilometers/min
	0.0098682	knots
	0.3048	meters/min
	0.00508	meters/sec
	0.011364	miles/hr
	1.8934×10^{-4}	miles/min
feet/sec	30.480	centimeters/sec
	3600.	feet/hr
	60.	feet/min
	1.0973	kilometers/hr
	0.018289	kilometers/min
	0.5921	knots
	18.29	meters/min
	0.3048	meters/sec
	0.6818	miles/hr
	0.011364	miles/min
feet/sec^2	1.	celos
	30.480	centimeters/sec^2
	0.0310810	g's (gravity)
	12.	inches/sec^2
	1.09728	kilometers/hr-sec
	0.30480	meters/sec^2
	40.9090	miles/hr-min
	0.6818	miles/hr-sec
fermis	1.0×10^{-5}	angstroms
	3.28084×10^{-15}	feet
	3.9370×10^{-14}	inches
	1.0×10^{-15}	meters
	1.	yukawas

MULTIPLY	BY	TO OBTAIN
ferrados	7.716	square feet
	0.7168	square meters
	1.	square varas
firkins	0.28571	barrels (31.5 U.S. gallons)
	1.2031	cubic feet
	2079.0	cubic inches
	0.034069	cubic meters
	0.044560	cubic yards
	9216.	drams
	9.	gallons (U.S., liquid)
	288.	gills
	0.14286	hogsheads
	34.069	liters
	1152.	ounces (U.S., liquid)
	72.	pints (U.S., liquid)
	36.	quarts (U.S., liquid)
fissions	200	MeV (total)
	8.9×10^{-18}	kilowatt-hr (total)
foot-candles	0.0010764	lumens/cm^2
	1.	lumens/ft^2
	0.0069444	lumens/in^2
	10.764	lumens/m^2
	10.764	lux
	10.764	meter-candles
	1.0764	milliphots
	10,764.	nox
	0.0010764	phots
foot-grains	1937.1	ergs
	1.4286×10^{-4}	foot-lbf
	1.9757	gram-cm
	1.937×10^{-4}	joules
foot-grains/sec	1.937×10^{-4}	watts
foot-kips	1.356	kilonewton-m
foot-lamberts	10.764	apostilbs
	0.31831	candles/ft^2
	0.0022105	candles/in^2
	3.4263	candles/m^2
	0.001076	lamberts
	1.0764	millilamberts
	3.426×10^{-4}	stilbs

PROFESSIONAL PUBLICATIONS, INC. ● Belmont, CA

MULTIPLY	BY	TO OBTAIN
foot-lbf	0.001285	BTU
	13.381	cubic centimeter-atm
	4.7254×10^{-4}	cubic foot-atm
	1.3558×10^{7}	dyne-cm
	8.4623×10^{18}	electron volts
	1.3558×10^{7}	ergs
	32.174	foot-poundals
	0.32389	gram-cal
	13,825.5	gram-cm
	5.1206×10^{-7}	horsepower-hr (metric)
	5.0505×10^{-7}	horsepower-hr (U.S.)
	12.	inch-lbf
	1.35582	joules
	3.2389×10^{-4}	kilogram-cal
	0.138255	kilogram-m
	3.7662×10^{-7}	kilowatt-hr
	0.013381	liter-atm
	13.558	megalergs
	1.3558	newton-m
	3.7662×10^{-4}	watt-hr
	1.356	watt-sec
foot-lbf/hr	2.1404×10^{-5}	BTU/min
	3.567×10^{-7}	BTU/sec
	3766.6	dyne-cm/sec
	3766.6	ergs/sec
	0.016667	foot-lbf/min
	2.7778×10^{-4}	foot-lbf/sec
	3.8409	gram-cm/sec
	5.1212×10^{-7}	horsepower (metric)
	5.05×10^{-7}	horsepower (U.S.)
	3.7666×10^{-4}	joules/sec
	5.3981×10^{-6}	kilogram-cal/min
	8.9970×10^{-8}	kilogram-cal/sec
	3.8409×10^{-5}	kilogram-m/sec
	3.7667×10^{-7}	kilowatts
	0.25175	lumens
	3.7666×10^{-4}	watts
foot-lbf/in^2	2.101×10^{3}	joules/m^2
foot-lbf/lbm	2.98907	joules/kg
foot-lbf/lbm-°R	5.3803	joules/kg·K

MULTIPLY	BY	TO OBTAIN
foot-lbf/min	0.0012841	BTU/min
	2.1402×10^{-5}	BTU/sec
	225,970.	dyne-cm/sec
	225,970.	ergs/sec
	60.	foot-lbf/hr
	0.016667	foot-lbf/sec
	230.43	gram-cm/sec
	3.072×10^{-5}	horsepower (metric)
	3.030×10^{-5}	horsepower (U.S.)
	0.022597	joules/sec
	3.2389×10^{-4}	kilogram-cal/min
	5.3980×10^{-6}	kilogram-cal/sec
	0.0023043	kilogram-m/sec
	2.2597×10^{-5}	kilowatts
	15.104	lumens
	0.022597	watts
foot-lbf/sec	4.6227	BTU/hr
	0.077045	BTU/min
	0.0012841	BTU/sec
	1.3558×10^{7}	dyne-cm/sec
	1.3558×10^{7}	ergs/sec
	3600.	foot-lbf/hr
	60.	foot-lbf/min
	13,826.	gram-cm/sec
	0.0018434	horsepower (metric)
	0.0018182	horsepower (U.S.)
	1.35582	joules/sec
	0.019433	kilogram-cal/min
	3.2388×10^{-4}	kilogram-cal/sec
	0.13826	kilogram-m/sec
	0.0013558	kilowatts
	906.272	lumens
	1.3558	watts
foot-poundals	3.994×10^{-5}	BTU
	0.4159	cubic centimeter-atm
	1.469×10^{-5}	cubic foot-atm
	421,401.	dyne-cm
	421,401.	ergs
	0.031081	foot-lbf
	0.010067	gram-cal
	429.7	gram-cm
	1.570×10^{-8}	horsepower-hr (U.S.)
	0.042140	joules
	1.007×10^{-5}	kilogram-cal
	0.004297	kilogram-m
	1.171×10^{-8}	kilowatt-hr
	4.1589×10^{-4}	liter-atm
	0.42141	megalergs
	1.170×10^{-5}	watt-hr

MULTIPLY	BY	TO OBTAIN
fortnights	14.	days
	2.	weeks
furlongs	10.	chains (Gunter's or surveyor's)
	6.6	chains (Ramsden's or engineer's)
	660.	feet
	0.201168	kilometers
	1000.	links (Gunter's or surveyor's)
	660.	links (Ramsden's or engineer's)
	201.168	meters
	0.125	miles
	264.	paces
	40.	rods (also perches or poles)
	220.	yards

PROFESSIONAL PUBLICATIONS, INC. ● Belmont, CA

MULTIPLY	BY	TO OBTAIN
g's (accel. due to gravity)	980.665	centimeters/sec^2
	32.174	feet/sec^2
	386.089	inches/sec^2
	35.3039	kilometers/hr-sec
	9.80665	meters/sec^2
	1316.21	miles/hr-min
	21.9369	miles/hr-sec
	32.174	pound (mass)-ft/lbf-sec^2
galileos	1.	centimeters/sec^2
	0.03281	feet/sec^2
	0.0010197	g (gravity)
	0.036	kilometers/hr-sec
	0.01	meters/sec^2
	1.3422	miles/hr-min
	0.02237	miles/hr-sec
gallons (Imperial)	1.2	gallons (U.S., liquid)
gallons (U.S., dry)	0.125	bushels
	4405.	cubic centimeters
	0.15556	cubic feet
	268.80	cubic inches
	0.00576	cubic yards
	1.16365	gallons (U.S., liquid)
	4.4049	liters
	0.5	pecks
	9.309	pints (U.S., liquid)
	4.	quarts (U.S., dry)
	4.6546	quarts (U.S., liquid)
gallons (U.S., liquid)	3.0689×10^{-6}	acre-ft
	0.031746	barrels (31.5 U.S. gallons)
	0.0238095	barrels (42 U.S. gallons)
	3785.4	cubic centimeters
	0.13368	cubic feet
	231.	cubic inches
	0.0037854	cubic meters
	0.004951	cubic yards
	1024.	drams
	0.111111	firkins
	0.859367	gallons (U.S., dry)
	32.	gills
	0.015873	hogsheads
	3.7854	liters
	3785.4	milliliters
	61,440.	minims
	128.	ounces (U.S., liquid)
	8.	pints (U.S., liquid)
	3.4375	quarts (U.S., dry)
	4.	quarts (U.S., liquid)

MULTIPLY	BY	TO OBTAIN
gallons of H_2O	8.345	pounds of H_2O
gallons/acre-day (gad)	2.296×10^{-5}	gallons/ft^2-day
gallons/day (gpd)	9.283×10^{-5}	cubic feet/min
	1.5472×10^{-6}	cubic feet/sec
	2.6288×10^{-6}	cubic meters/min
	0.002629	liters/min
	1.0×10^{-6}	millions of gallons/day
gallons/day-ft^2	1.	**Meinzer units**
	43,560.	**gallons/day-acre**
	0.04356	**millions of gallons/day-acre**
gallons/hr	0.1337	cubic feet/hr
	0.002228	cubic feet/min
	3.71×10^{-5}	cubic feet/sec
	6.309×10^{-5}	cubic meters/min
	0.016667	gallons (U.S., liquid)/min
	2.7778×10^{-4}	gallons (U.S., liquid)/sec
	0.06309	liters/min
	0.001052	liters/sec
gallons/min	34.286	barrels (42 U.S. gallons)/day
	1.4286	barrels (42 U.S. gallons)/hr
	0.02381	barrels (42 U.S. gallons)/min
	63.1	cubic centimeters/sec
	192.5	cubic feet/day
	8.02	cubic feet/hr
	0.1337	cubic feet/min
	0.0022280	cubic feet/sec
	0.0049512	cubic yards/min
	1440.	gallons/day
	0.0166667	gallons (U.S., liquid)/sec
	3.7854	liters/min
	0.06309	liters/sec
	0.11139	miner's inches*
	8.345	pounds H_2O/min
gallons/sec	3784.7	cubic centimeters/sec
	481.3	cubic feet/hr
	8.0208	cubic feet/min
	0.1337	cubic feet/sec
	0.2971	cubic yards/min
	60.	gallons (U.S., liquid)/min
	227.13	liters/min
	3.7854	liters/sec
	6.6828	miner's inches*
	500.6	pounds H_2O/min

---

* ID, KS, ND, NE, NM, NV, SD, UT

MULTIPLY	BY	TO OBTAIN
gammas	7.958×10^{-6}	ampere-turns/cm
	2.0213×10^{-5}	ampere-turns/in
	1.0×10^{-5}	gauss
	1.0×10^{-5}	gilberts/cm
	1.0×10^{-5}	lines/cm^2
	6.452×10^{-5}	lines/in^2
	1.0×10^{-9}	teslas
gauss	0.7958	ampere-turns/cm
	2.0213	ampere-turns/in
	100,000.	gammas
	1.	gilberts/cm
	1.	lines/cm^2
	6.452	lines/in^2
	1.	maxwells/cm^2
	6.4516	maxwells/in^2
	3.3359×10^{-11}	statwebers
	1.0×10^{-4}	teslas
	1.0×10^{-8}	webers/cm^2
	6.452×10^{-8}	webers/in^2
	1.0×10^{-4}	webers/m^2
geepounds	14.594	kilograms
	1.	pounds (force)-sec^2/ft
	32.174	pounds (mass)
	1.	slugs
GeV	1.0×10^{12}	electron volts
gilberts	0.07958	abampere-turns
	0.7958	ampere-turns
gilberts/cm	0.7958	ampere-turns/cm
	2.0213	ampere-turns/in
	100,000.	gammas
	1.	gauss
	1.	lines/cm^2
	6.452	lines/in^2
	1.0×10^{-4}	teslas

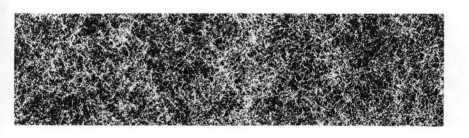

PROFESSIONAL PUBLICATIONS, INC. • Belmont, CA

MULTIPLY	BY	TO OBTAIN
gills (U.S.)	9.9206×10^{-4}	barrels (31.5 U.S. gallons)
	118.294	cubic centimeters
	0.0041775	cubic feet
	7.21875	cubic inches
	1.1829×10^{-4}	cubic meters
	1.54723×10^{-4}	cubic yards
	32.	drams
	0.0034722	firkins
	0.03125	gallons (U.S., liquid)
	4.9603×10^{-4}	hogsheads
	0.118294	liters
	118.29	milliliters
	1920.	minims
	4.	ounces (U.S., liquid)
	0.25	pints (U.S., liquid)
	0.125	quarts (U.S., liquid)
grads (grades)	0.9	degrees
	54.	minutes
	0.01	quadrants
	0.015708	radians
	0.0025	revolutions
	3240.	seconds
	0.03	signs
grains	0.324	carats
	6.4799	centigrams
	0.036571	drams (avoir)
	0.0166667	drams (troy)
	63.546	dynes
	0.064799	grams
	6.47989×10^{-5}	kilograms
	6.3546×10^{-4}	newtons
	0.0022857	ounces (avoir)
	0.0020833	ounces (troy)
	0.0416667	pennyweights
	1.4286×10^{-4}	pounds (avoir)
	1.7361×10^{-4}	pounds (troy)
	0.0045963	poundals
	0.05	scruples
	7.1429×10^{-8}	tons (short)

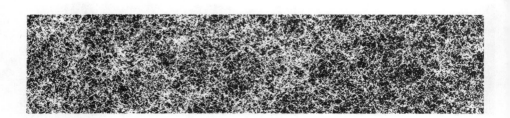

PROFESSIONAL PUBLICATIONS, INC. ● Belmont, CA

MULTIPLY	BY	TO OBTAIN
grains/ft^3	2.2883×10^{-6}	grams/cm^3
	2.2884	grams/m^3
	2.2883×10^{-6}	grams/ml
	0.0022884	kilograms/m^3
	1.68314×10^{-7}	mercury at 0°C
	6.480×10^{-5}	milligrams/ft^3
	2288.35	milligrams/m^3
	1.42857×10^{-4}	pounds/ft^3
	1.90959×10^{-5}	pounds/gallon
	8.26744×10^{-8}	pounds/in^3
	7.791486×10^{-13}	pounds/circular mil-ft
	4.43975×10^{-6}	slugs/ft^3
grains/gallon (U.S., liquid)	1.2	degrees, Clarke
	0.0078125	grains/ounce (U.S., liquid)
	17.118	grams/m^3
	17.118	milligrams/liter (mg/ℓ)
	0.0022857	ounces/gallon (U.S., liquid)
	17.118	parts per million (ppm)
	0.0010687	pounds/ft^3
	142.9	pounds/million gallons
grains/in	0.02551	grams/cm
	2.5511	kilograms/km
	0.0025511	kilograms/m
	0.0017143	pounds/ft
	9.0514	pounds/mi
	0.005143	pounds/yd
	0.004526	tons (short)/mi
grains/ounce (U.S., liquid)	0.036571	drams/ounce (U.S., liquid)
	128.	grains/gallon (U.S., liquid)
	2.1911	kilograms/m^3
	0.29257	ounces/gallon (U.S., liquid)
	0.0012665	ounces (U.S., liquid)/in^3
	0.073143	ounces/quart (U.S., liquid)
	0.13679	pounds/ft^3
	7.9159×10^{-5}	pounds/in^3

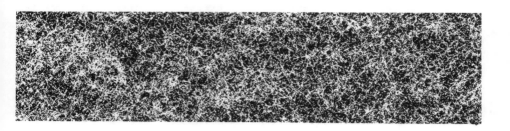

PROFESSIONAL PUBLICATIONS, INC. ● Belmont, CA

MULTIPLY	BY	TO OBTAIN
grams	5.	carats
	100.	centigrams
	6.0232×10^{23}	daltons
	0.1	decagrams
	10.	decigrams
	0.56438	drams (avoir)
	0.25721	drams (troy)
	980.665	dynes
	15.432	grains
	0.001	kilograms
	1.0×10^6	micrograms
	1000.	milligrams
	0.00980665	newtons
	0.035274	ounces (avoir)
	0.032151	ounces (troy)
	0.64302	pennyweights
	0.0022046	pounds (avoir)
	0.0026792	pounds (troy)
	0.070932	poundals
	0.77162	scruples
	6.8522×10^{-5}	slugs
	0.034286	tons (assay)
	9.8421×10^{-7}	tons (long)
	1.0×10^{-6}	tons (metric)
	1.1023×10^{-6}	tons (short)
grams/cm^2	9.678×10^{-4}	atmospheres
	9.8067×10^{-4}	bars
	0.073556	centimeters Hg at 0°C
	1.0	centimeters H$_2$O at 4°C
	980.665	dynes/cm^2
	0.028959	inches Hg at 32°F
	0.001	kilograms/cm^2
	10.	kilograms/m^2
	1.0×10^{-5}	kilograms/mm^2
	0.22757	ounces/in^2
	2.0482	pounds/ft^2
	0.01422	pounds/in^2
	0.0010241	tons/ft^2
	7.112×10^{-6}	tons/in^2

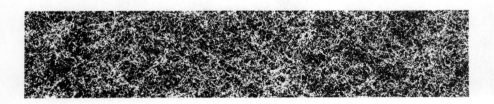

PROFESSIONAL PUBLICATIONS, INC. ● Belmont, CA

MULTIPLY	BY	TO OBTAIN
grams/cm^3	980.665	dynes/cm^3
	436,996.	grains/ft^3
	1.0×10^6	grams/m^3
	1.0	grams/ml
	1000.	kilograms/m^3
	62.428	pounds/ft^3
	8.3454	pounds/gal
	0.03613	pounds/in^3
	3.405×10^{-7}	pounds/mil ft
	1.94032	slugs/ft^3
grams/cm-sec	100.	centipoises
	1.	dyne-sec/cm^2
	0.010197	kilogram (force)-sec/m^2
	360.	kilograms/m-hr
	1.	poises
	241.93	pounds (mass)/ft-hr
	0.067204	pounds (mass)/ft-sec
	0.002089	pound-sec/ft^2
	1.450×10^{-5}	pound-sec/in^2
	241.93	poundal-hr/ft^2
	0.067204	poundal-sec/ft^2
grams/ℓ	0.016691	drams/ounce (U.S., liquid)
	58.418	grains/gallon (U.S., liquid)
	0.45639	grains/ounce (U.S., liquid)
	1000.	grams/m^3
	1.	kilograms/m^3
	1000.	micrograms/ml
	1000.	milligrams/ℓ
	0.1335	ounces (U.S., liquid)/gallon (U.S., liquid)
	0.033382	ounces (U.S., liquid)/quart (U.S., liquid)
	2719.4	pounds/acre-ft
	0.062428	pounds/ft^3
	3.6127×10^{-5}	pounds/in^3
	0.001	tons (metric)/m^3
	4.5949×10^6	tons (short)/mi^3

PROFESSIONAL PUBLICATIONS, INC. • Belmont, CA

MULTIPLY	BY	TO OBTAIN
grams/m^3	1.6691×10^{-5}	drams/ounce (U.S., liquid)
	4.5639×10^{-4}	grains/ounce (U.S., liquid)
	0.437	grains/ft^3
	0.05842	grains/gallon (U.S., liquid)
	1.0×10^{-6}	grams/cm^3
	0.001	grams/ℓ
	1.0×10^{-6}	grams/ml
	0.001	kilograms/m^3
	1.	micrograms/ml
	1.	milligrams/ℓ
	3.3382×10^{-5}	ounces/quart (U.S., liquid)
	6.2428×10^{-5}	pounds/ft^3
	8.34513×10^{-6}	pounds/gallon (U.S., liquid)
	3.613×10^{-8}	pounds/in^3
	3.40497×10^{-13}	pounds/mil ft
	1.94022×10^{-6}	slugs/ft^3
grams/ml	436,996.	grains/ft^3
	1.0	grams/cm^3
	1.0×10^{6}	grams/m^3
	1000.	kilograms/m^3
	62.4280	pounds/ft^3
	8.34540	pounds/gallon (U.S., liquid)
	0.036127	pounds/in^3
	3.40495×10^{-7}	pounds/mil ft
	1.94032	slugs/ft^3
grams/ton (long)	0.9842	grams/ton (metric)
	0.96207	grams/ton (short)
	2.109×10^{-4}	karats
	9.842×10^{-4}	milligrams/g
	0.9842	milligrams/kg
	0.028714	milligrams/ton (assay)
	0.035274	ounces/ton (long)
	0.031494	ounces/ton (short)
	0.0022046	pounds/ton (long)
	0.0019684	pounds/ton (short)

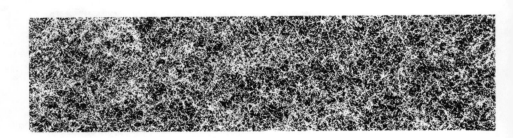

PROFESSIONAL PUBLICATIONS, INC. • Belmont, CA

MULTIPLY	BY	TO OBTAIN
grams/ton (metric)	1.01605	grams/ton (long)
	0.977517	grams/ton (short)
	2.14286×10^{-4}	karats
	0.001	milligrams/g
	1.	milligrams/kg
	0.029175	milligrams/ton (assay)
	0.035840	ounces/ton (long)
	0.032	ounces/ton (short)
	0.0022400	pounds/ton (long)
	0.002	pounds/ton (short)
grams/ton (short)	1.0394	grams/ton (long)
	1.023	grams/ton (metric)
	2.19214×10^{-4}	karats
	0.001023	milligrams/g
	1.023	milligrams/kg
	0.029846	milligrams/ton (assay)
	0.036643	ounces/ton (long)
	0.032736	ounces/ton (short)
	0.00229152	pounds/ton (long)
	0.002046	pounds/ton (short)
gram-cal (see calories)	0.003969	BTU
	0.0022	CHU
	41.312	cubic centimeter-atm
	0.001459	cubic foot-atm
	4.186×10^{7}	dyne-cm
	4.186×10^{7}	ergs
	3.0874	foot-lbf
	99.334	foot-poundals
	42,686.	gram-cm
	1.559×10^{-6}	horsepower-hr (U.S.)
	4.186	joules
	0.001	kilogram-cal
	0.42685	kilogram-m
	1.1628×10^{-6}	kilowatt-hr
	0.041312	liter-atm
	41.861	megalergs
	0.0011628	watt-hr
	4.18605	watt-sec
gram-cal/cm^2-hr	88.48	BTU/ft^2-day
	3.687	BTU/ft^2-hr
	2.778×10^{-4}	gram-cal/cm^2-sec
	0.001163	watts/cm^2

PROFESSIONAL PUBLICATIONS, INC. • Belmont, CA

MULTIPLY	BY	TO OBTAIN
gram-cal/cm^2-hr-$^\circ$C	49.16	BTU/ft^2-day-$^\circ$F
	2.048	BTU/ft^2-hr-$^\circ$F
	2.7778×10^{-4}	gram-cal/cm^2-sec-$^\circ$C
	0.001163	watts/cm^2-$^\circ$C
gram-cal/cm^2-sec	318,531.	BTU/ft^2-day
	13,272.	BTU/ft^2-hr
	3600.	gram-cal/cm^2-hr
	4.187	watts/cm^2
gram-cal/cm^2-sec-$^\circ$C	176,962.	BTU/ft^2-day-$^\circ$F
	7373.	BTU/ft^2-hr-$^\circ$F
	3600.	gram-cal/cm^2-hr-$^\circ$C
	4.187	watts/cm^2-$^\circ$C
gram-cal-cm/cm^2-hr-$^\circ$C	0.0672	BTU-ft/ft^2-hr-$^\circ$F
	19.35	BTU-in/ft^2-day-$^\circ$F
	0.8058	BTU-in/ft^2-hr-$^\circ$F
	2.7778×10^{-4}	gram-cal-cm/cm^2-sec-$^\circ$C
	0.001163	watts-cm/cm^2-$^\circ$C
gram-cal-cm/cm^2-sec-$^\circ$C	241.9	BTU-ft/ft^2-hr-$^\circ$F
	69,670.	BTU-in/ft^2-day-$^\circ$F
	2903.	BTU-in/ft^2-hr-$^\circ$F
	3600.	gram-cal-cm/cm^2-hr-$^\circ$C
	4.186	joules-cm/sec-cm^2-$^\circ$C
	850.6	joules-in/sec-ft^2-$^\circ$F
	4.186	watts-cm/cm^2-$^\circ$C
gram-cal/g	1.8	BTU/lbm
gram-cal/min	0.0039683	BTU/min
	3.086	foot-lbf/min
	0.00514	foot-lbf/sec
	9.487×10^{-5}	horsepower (metric)
	9.356×10^{-5}	horsepower (U.S.)
	0.069767	watts
gram-cal/sec	0.0039683	BTU/sec
	3.0875	foot-lbf/sec
	0.005692	horsepower (metric)
	0.0056136	horsepower (U.S.)
	4.186	watts

MULTIPLY	BY	TO OBTAIN
gram-cm	9.295×10^{-8}	BTU
	9.678×10^{-4}	cubic centimeter-atm
	3.418×10^{-8}	cubic foot-atm
	980.7	dyne-cm
	980.7	ergs
	7.233×10^{-5}	foot-lbf
	0.002327	foot-poundals
	2.343×10^{-5}	gram-cal
	3.653×10^{-11}	horsepower-hr (U.S.)
	9.8067×10^{-5}	joules
	2.343×10^{-8}	kilogram-cal
	1.0×10^{-5}	kilogram-m
	2.724×10^{-11}	kilowatt-hr
	9.678×10^{-7}	liter-atm
	9.8067×10^{-4}	megalergs
	9.8067×10^{-5}	newton-m
	2.724×10^{-8}	watt-hr
	9.80665×10^{-5}	watt-sec
gram-cm^2	0.001	kilogram-cm^2
	1.0×10^{-7}	kilogram-m^2
	2.373×10^{-6}	pound-ft^2
	3.4172×10^{-4}	pound-in^2
gram-cm/sec	5.573×10^{-6}	BTU/min
	9.288×10^{-8}	BTU/sec
	980.665	dyne-cm/sec
	980.665	ergs/sec
	0.2604	foot-lbf/hr
	0.004340	foot-lbf/min
	7.233×10^{-5}	foot-lbf/sec
	1.3333×10^{-7}	horsepower (metric)
	1.3151×10^{-7}	horsepower (U.S.)
	9.80665×10^{-5}	joules/sec
	1.4056×10^{-6}	kilogram-cal/min
	2.343×10^{-8}	kilogram-cal/sec
	1.0×10^{-5}	kilogram-m/sec
	9.80665×10^{-8}	kilowatts
	0.06555	lumens
	9.8067×10^{-5}	watts
gram-sec/cm^2	980.665	poises
	2.0482	pound-sec/ft^2
	0.014223	pound-sec/in^2
grays	100.	rads
gross	12.	dozens
	144.	units

PROFESSIONAL PUBLICATIONS, INC. ● Belmont, CA

MULTIPLY	BY	TO OBTAIN
hands	10.16	centimeters
	5.333	digits
	0.333333	feet
	4.	inches
	0.1016	meters
	0.133333	paces
	1.333	palms
	0.4444	spans
	0.111111	yards
hectares	2.47104	acres
	100.	ares
	10,000.	centares
	9.8842	roods
	0.003861	sections
	24.7104	square chains (Gunter's or surveyor's)
	107,639.	square feet
	0.01	square kilometers
	247.104	square links (Gunter's or surveyor's)
	10,000.	square meters
	0.00386102	square miles
	395.369	square rods (also square perches)
	11,959.9	square yards
	1.07250×10^{-4}	townships
hectograms	100.	grams
	0.1	kilograms
	3.5274	ounces (avoir)
	0.22046	pounds (avoir)
hectoliters	3.532	cubic feet
	26.42	gallons (U.S., liquid)
	100.	liters
hectometers	328.1	feet
	0.1	kilometers
	100.	meters
hefners	0.9	bougie decimales
	0.9	candles (int'l)
	0.0937	carcel units
	0.9	English sperm candles
	0.9	lumens/sterad
	0.9	pentane candles
hemispheres	0.5	spheres
	360.	spherical degrees
	4.	spherical right angles
	20,626.5	square degrees
	6.28319	steradians
	0.5	steregons

MULTIPLY	BY	TO OBTAIN
henrys	1.0×10^9	abhenrys
	1.0×10^6	microhenrys
	1000.	millihenrys
	1.	ohm-sec
	1.	volt-sec/amp
	1.	webers/ampere
henrys (absolute)	1.11265×10^{-12}	stathenrys
hertz	1.	cycles/sec
hogsheads	2.	barrels (31.5 U.S. gallons)
	8.4219	cubic feet
	14,553.	cubic inches
	0.23848	cubic meters
	0.31192	cubic yards
	7.	firkins
	63.	gallons (U.S., liquid)
	2016.	gills
	0.23848	kiloliters
	238.48	liters
	8064.	ounces (U.S., liquid)
	504.	pints (U.S., liquid)
	252.	quarts (U.S., liquid)
horsepower (boiler)	33,446.	BTU/hr
	9.2905	BTU/sec
	9809.5	watts
horsepower (continental)	41.84	BTU/min
	75.	kilogram-m/sec
	736.	watts
horsepower (electrical)	0.7065	BTU/sec
	7.46×10^9	ergs/sec
	550.22	foot-lbf/sec
	178.211	gram-cal/sec
	1.000402	horsepower (U.S.)
	746.	watts
horsepower (mechanical)	2545.0	BTU/hr
	550.	foot-lbf/sec
	0.99960	horsepower (electric)
	1.	horsepower (U.S.)
	0.7457	kilowatts
	745.7	watts

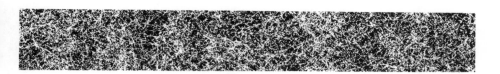

PROFESSIONAL PUBLICATIONS, INC. ● Belmont, CA

MULTIPLY	BY	TO OBTAIN
horsepower (metric)	41.795	BTU/min
	0.69658	BTU/sec
	7.355×10^9	dyne-cm/sec
	7.355×10^9	ergs/sec
	1.953×10^6	foot-lbf/hr
	32,549.	foot-lbf/min
	542.48	foot-lbf/sec
	7.50×10^6	gram-cm/sec
	0.98632	horsepower (U.S.)
	735.50	joules/sec
	10.542	kilogram-cal/min
	0.1757	kilogram-cal/sec
	75.	kilogram-m/sec
	0.73550	kilowatts
	491,640.	lumens
	735.50	watts
horsepower (U.S.)	2542.5	BTU/hr
	42.375	BTU/min
	0.70624	BTU/sec
	7.4570×10^9	dyne-cm/sec
	7.4570×10^9	ergs/sec
	1.980×10^6	foot-lbf/hr
	33,000.	foot-lbf/min
	550.	foot-lbf/sec
	7.604×10^6	gram-cm/sec
	0.999598	horsepower (electric)
	1.	horsepower (mechanical)
	1.0139	horsepower (metric)
	745.7	joules/sec
	10.69	kilogram-cal/min
	0.17811	kilogram-cal/sec
	76.040	kilogram-m/sec
	0.7457	kilowatts
	498,433.	lumens
	745.7	watts
horsepower (water)	42.394	BTU/min
	746.04	watts
horsepower-hr (metric)	2509.8	BTU
	1.953×10^6	foot-lbf
	0.98632	horsepower-hr (U.S.)
	2.6476×10^6	joules
	632.467	kilogram-cal
	2.70×10^5	kilogram-m
	735.5	watt-hr

PROFESSIONAL PUBLICATIONS, INC. • Belmont, CA

MULTIPLY	BY	TO OBTAIN
horsepower-hr (U.S.)	2542.47	BTU
	2.649×10^7	cubic centimeter-atm
	935.6	cubic foot-atm
	2.685×10^{13}	dyne-cm
	2.685×10^{13}	ergs
	1.98×10^6	foot-lbf
	6.370×10^7	foot-poundals
	641,300.	gram-cal
	2.737×10^{10}	gram-cm
	1.01387	horsepower-hr (metric)
	2.6845×10^6	joules
	641.2	kilogram-cal
	273,745.	kilogram-m
	0.7457	kilowatt-hr
	26,494.	liter-atm
	2.6845×10^7	megalergs
	745.7	watt-hr
hours (solar)	0.041781	days (sidereal)
	0.041667	days (solar)
	60.	minutes (solar)
	0.0014110	months (lunar)
	0.0013699	months (mean)
	3609.9	seconds (sidereal)
	3600.	seconds (solar)
	0.0059524	weeks
	1.1384×10^{-4}	years (leap)
	1.14075×10^{-4}	years (sidereal)
	1.14080×10^{-4}	years (solar)
hundredweights (long)	1.12	centals
	1.12	hundredweights (short)
	50.802	kilograms
	1792.	ounces (avoir)
	112.	pounds (avoir)
	0.2	quarters (long)
	0.224	quarters (short)
	0.50802	quintals
	3.4811	slugs
	0.05	tons (long)
	0.050802	tons (metric)
	0.056	tons (short)

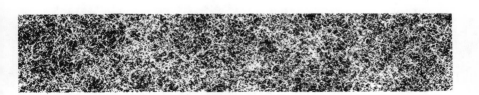

PROFESSIONAL PUBLICATIONS, INC. • Belmont, CA

MULTIPLY	BY	TO OBTAIN
hundredweights (short)	1.	centals
	0.89286	hundredweights (long)
	45.359	kilograms
	1600.	ounces (avoir)
	100.	pounds (avoir)
	0.17857	quarters (long)
	0.2	quarters (short)
	0.45359	quintals
	3.1081	slugs
	0.044643	tons (long)
	0.045359	tons (metric)
	0.05	tons (short)

MULTIPLY	BY	TO OBTAIN
inches	2.54×10^8	angstroms
	3.	barleycorns
	100.	calibers
	2.54	centimeters
	0.0012626	chains (Gunter's or surveyor's)
	8.3333×10^{-4}	chains (Ramsden's or engineer's)
	0.0555556	cubits
	0.00254	decameters
	0.254	decimeters
	1.3333	digits
	144.	douziemes
	0.022222	ells
	6.023	ems (pica)
	0.0138889	fathoms
	0.0833333	feet
	2.54×10^{13}	fermis
	0.25	hands
	2.540×10^{-5}	kilometers
	12.	lines
	0.126263	links (Gunter's or surveyor's)
	0.0833333	links (Ramsden's or engineer's)
	0.0254	meters
	2.540×10^4	micrometers
	2.54×10^{10}	micromicrons
	25,400.	microns
	1.5783×10^{-5}	miles (statute)
	25.40	millimeters
	2.54×10^7	millimicrons
	2.54×10^{10}	millionth microns
	1000.	mils
	0.4444	nails
	2.54×10^7	nanometers
	0.0333	paces
	0.333	palms
	0.0050505	perches
	72.27	points
	0.0050505	rods (also perches or poles)
	0.11111	spans
	2.54×10^{10}	stigmas
	2.54×10^8	tenthmeters
	0.03	varas
	39,450.4	wavelengths red line cadmium
	2.54×10^{11}	X-units
	0.0277778	yards
	2.54×10^{13}	yukawas

PROFESSIONAL PUBLICATIONS, INC. • Belmont, CA

MULTIPLY	BY	TO OBTAIN
inches Hg at 32°F	0.03342	atmospheres
	0.03386	bars
	2.54	centimeters Hg at 0°C
	33,864.	dynes/cm^2
	34.532	grams/cm^2
	13.6	inches H$_2$O
	0.034532	kilograms/cm^2
	345.32	kilograms/m^2
	3.4532×10^{-4}	kilograms/mm^2
	7.8582	ounces/in^2
	3386.4	pascals
	70.726	pounds/ft^2
	0.49115	pounds/in^2
	0.035363	tons (short)/ft^2
	2.4558×10^{-4}	tons (short)/in^2
inches of run-off	53.3	acre-ft/mile^2
	2.323×10^6	cubic feet/mile^2
inches of water	0.0735	inches Hg
	248.6	pascals
	5.199	pounds/ft^2
	0.0361	pounds/in^2
inches/sec^2	2.54	centimeters/sec^2
	0.083333	feet/sec^2
	0.0025901	g (gravity)
	0.091440	kilometers/hr-sec
	0.0254	meters/sec^2
	3.4091	miles/hr-min
	0.056818	miles/hr-sec
inch-lbf	1.07×10^{-4}	BTU
	0.08333	foot-lbf
	0.1130	joules
	4.21×10^{-8}	horsepower-hr (U.S.)
	3.137×10^{-8}	kilowatt-hr
	0.1130	newton-m

PROFESSIONAL PUBLICATIONS, INC. • Belmont, CA

MULTIPLY	BY	TO OBTAIN
joules	6.705×10^9	amu (nuclear)
	9.471×10^{-4}	BTU
	0.238846	calories (IST)
	0.238662	calories (mean)
	0.239006	calories (thermo)
	9.8692	cubic centimeter-atm
	3.485×10^{-4}	cubic foot-atm
	1.0×10^7	dyne-cm
	6.2415×10^{18}	electron volts
	1.0×10^7	ergs
	0.73756	foot-lbf
	23.73	foot-poundals
	0.23889	gram-cal
	10,197.	gram-cm
	3.725×10^{-7}	horsepower-hr (U.S.)
	8.8495	inch-lbf
	2.388×10^{-4}	kilogram-cal
	0.10197	kilogram-m
	1.113×10^{-17}	kilograms (nuclear)
	2.7778×10^{-7}	kilowatt-hr
	0.009869	liter-atm
	10.	megalergs
	0.10197	meter-kilogram (force)
	1.	newton-m
	9.478×10^{-9}	therms
	2.7778×10^{-4}	watt-hr
	1.	watt-sec
joules/cm	100.	newtons
joules/g-°C	0.238846	BTU/lbm-°F
joules/kg	4.2992×10^{-4}	BTU/pound
joules/kg-°C	238.846	BTU/lbm-°F
joules/m	1.	newtons
joules/m²	2.390×10^{-5}	langleys
joules/m³	2.6839×10^{-5}	BTU/ft³
joules/m³-°C	1.4911×10^{-5}	BTU/ft³-°F

PROFESSIONAL PUBLICATIONS, INC. • Belmont, CA

MULTIPLY	BY	TO OBTAIN
joules/sec	0.056869	BTU/min
	9.4826×10^{-4}	BTU/sec
	1.0×10^{7}	dyne-cm/sec
	1.0×10^{7}	ergs/sec
	2654.9	foot-lbf/hr
	44.254	foot-lbf/min
	0.73756	foot-lbf/sec
	10,197.2	gram-cm/sec
	0.0013596	horsepower (metric)
	0.001341	horsepower (U.S.)
	0.014333	kilogram-cal/min
	2.3885×10^{-4}	kilogram-cal/sec
	0.10197	kilogram-m/sec
	0.001	kilowatts
	668.45	lumens
	1.	watts

MULTIPLY	BY	TO OBTAIN
karats	4741.6	grams/ton (long)
	4666.67	grams/ton (metric)
	4561.7	grams/ton (short)
	4.66667	milligrams/g
	4666.67	milligrams/kg
	136.15	milligrams/ton (assay)
	167.25	ounces/ton (long)
	149.33	ounces/ton (short)
	10.4533	pounds/ton (long)
	9.3333	pounds/ton (short)
kilocalories (kcal)	3.968	BTU
	1000.	calories
	4.185×10^{10}	ergs
	3090.	foot-lbf
	1000.	gram-cal
	4190.	joules
	1.	kilogram-cal
kilocalories/hr-m-°C	0.672	BTU-ft/ft^2-hr-°F
kilocalories/hr-m^2-°C	0.2048	BTU/ft^2-hr-°F
kilocalories-cm/sec-cm^2-°C	3.60×10^6	gram-cal-cm/hr-cm^2-°C

MULTIPLY	BY	TO OBTAIN
kilograms	6.025×10^{26}	amu
	5000.	carats
	0.022046	centals
	100,000.	centigrams
	6.0232×10^{26}	daltons
	100.	decagrams
	10,000.	decigrams
	564.38	drams (avoir)
	257.206	drams (troy)
	5.610×10^{35}	electron volts (nuclear)
	15,432.	grains
	1000.	grams
	0.019684	hundredweights (long)
	0.022046	hundredweights (short)
	8.987×10^{16}	joules (nuclear)
	1.0×10^{9}	micrograms
	1.0×10^{6}	milligrams
	35.274	ounces (avoir)
	32.151	ounces (troy)
	643.02	pennyweights
	2.204623	pounds (avoir)
	2.67923	pounds (troy)
	0.0039369	quarters (long)
	0.0044093	quarters (short)
	0.01	quintals
	771.62	scruples
	0.068522	slugs
	0.1575	stones
	34.286	tons (assay)
	9.8421×10^{-4}	tons (long)
	0.001	tons (metric)
	0.0011023	tons (short)
kilograms (force)	98.0665	crinals
	980,665.	dynes
	1000.	grams (force)
	0.0022046	kips
	9.80665	newtons
	2.2046	pounds
	70.932	poundals
	0.00980665	sthènes

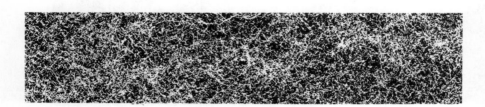

PROFESSIONAL PUBLICATIONS, INC. • Belmont, CA

MULTIPLY	BY	TO OBTAIN
kilograms (force)/cm^2	0.96783	atmospheres
	0.980665	bars
	73.556	centimeters Hg at 0°C
	1000.	centimeters H$_2$O at 4°C
	980,665.	dynes/cm^2
	1000.	grams/cm^2
	28.959	inches Hg at 32°F
	10,000.	kilograms/m^2
	0.01	kilograms/mm^2
	227.57	ounces/in^2
	2048.2	pounds/ft^2
	14.223	pounds/in^2
	0.91436	tons (long)/ft^2
	0.0063500	tons (long)/in^2
	1.0241	tons (short)/ft^2
	0.0071117	tons (short)/in^2
kilograms/day	41.666	grams/hr
	11.574	milligrams/sec
	1.470	ounces (avoir)/hr
	2.2046	pounds (avoir)/day
	0.09186	pounds (avoir)/hr
kilograms (force)/m^2	9.6784×10^{-5}	atmospheres
	9.8066×10^{-5}	bars
	0.0073556	centimeters Hg at 0°C
	0.1	centimeters H$_2$O at 4°C
	98.0665	dynes/cm^2
	0.0032809	feet of H$_2$O at 39.1°F
	0.1	grams/cm^2
	0.0028959	inches Hg at 0°C
	1.0×10^{-4}	kilograms/cm^2
	1.0×10^{-6}	kilograms/mm^2
	0.022757	ounces/in^2
	9.80665	pascals
	0.20482	pounds/ft^2
	0.0014223	pounds/in^2
	1.0241×10^{-4}	tons (short)/ft^2
	7.1117×10^{-7}	tons (short)/in^2

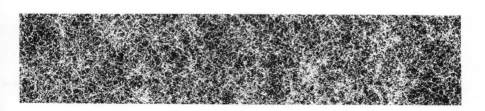

PROFESSIONAL PUBLICATIONS, INC. ● Belmont, CA

MULTIPLY	BY	TO OBTAIN
kilograms/m^3	0.016691	drams/ounce (U.S., liquid)
	437.	grains/ft^3
	0.45639	grains/ounce (U.S., liquid)
	0.001	grams/cm^3
	1.	grams/ℓ
	1000.	grams/m^3
	0.001	grams/ml
	7.35559×10^{-5}	mercury at 0°C
	1000.	micrograms/ml
	1000.	milligrams/ℓ
	0.13353	ounces/gallon (U.S., liquid)
	5.7804×10^{-4}	ounces (U.S., liquid)/in^3
	0.033382	ounces/quart (U.S., liquid)
	2719.4	pounds/acre-ft
	0.062428	pounds/ft^3
	3.6127×10^{-5}	pounds/in^3
	0.008345406	pounds/gallon (U.S., liquid)
	3.40500×10^{-10}	pounds/mil ft
	0.00194032	slugs/ft^3
	7.525×10^{-4}	tons (long)/yd^3
	0.001	tons (metric)/m^3
	4.5949×10^6	tons (short)/mi^3
kilograms/m-hr	0.27778	centipoises
	0.0027778	dyne-sec/cm^2
	0.0027778	grams/cm-sec
	2.84×10^{-5}	kilogram (force)-sec/m^2
	0.0027778	poises
	0.672	pounds (mass)/ft-hr
	1.86×10^{-4}	pounds (mass)/ft-sec
	5.79×10^{-6}	pound-sec/ft^2
	4.03×10^{-8}	pound-sec/in^2
	0.671	poundal-hr/ft^2
	1.86×10^{-4}	poundal-sec/ft^2
kilograms (force)/mm^2	96.782	atmospheres
	98.0665	bars
	7355.	centimeters Hg at 0°C
	9.80665×10^7	dynes/cm^2
	100,000.	grams/cm^2
	100.	kilograms/cm^2
	1.0×10^6	kilograms/m^2
	22,757.	ounces/in^2
	9.80665×10^6	pascals
	204,816.	pounds/ft^2
	1422.3	pounds/in^2
	102.41	tons (short)/ft^2
	0.71117	tons (short)/in^2

MULTIPLY	BY	TO OBTAIN
kilogram-cal	3.9683	BTU
	41,312.	cubic centimeter-atm
	1.459	cubic foot-atm
	4.186×10^{10}	dyne-cm
	4.186×10^{10}	ergs
	3087.4	foot-lbf
	99,334.	foot-poundals
	1000.	gram-cal
	4.2686×10^7	gram-cm
	0.001581	horsepower-hr (metric)
	0.001559	horsepower-hr (U.S.)
	4186.	joules
	1.	kilocalories
	426.85	kilogram-m
	0.001163	kilowatt-hr
	41.312	liter-atm
	41,860.	megalergs
	1.163	watt-hr
	4186.	watt-sec
kilogram-cal/kg	1.8	BTU/lbm
kilogram-cal/min	3.9695	BTU/min
	0.066158	BTU/sec
	6.9767×10^8	dyne-cm/sec
	6.9767×10^8	ergs/sec
	185,245.	foot-lbf/hr
	3087.5	foot-lbf/min
	51.457	foot-lbf/sec
	711,427.	gram-cm/sec
	0.094862	horsepower (metric)
	0.093559	horsepower (U.S.)
	69.767	joules/sec
	0.016667	kilogram-cal/sec
	7.1143	kilogram-m/sec
	0.069767	kilowatts
	46,636.	lumens
	69.767	watts

PROFESSIONAL PUBLICATIONS, INC. • Belmont, CA

MULTIPLY	BY	TO OBTAIN
kilogram-cal/sec	238.165	BTU/min
	3.9694	BTU/sec
	4.1860×10^{10}	dyne-cm/sec
	4.1860×10^{10}	ergs/sec
	1.111×10^7	foot-lbf/hr
	185,246.	foot-lbf/min
	3087.4	foot-lbf/sec
	4.2685×10^7	gram-cm/sec
	5.6917	horsepower (metric)
	5.6134	horsepower (U.S.)
	4186.	joules/sec
	60.	kilogram-cal/min
	426.85	kilogram-m/sec
	4.186	kilowatts
	2.798×10^6	lumens
	4186.	watts
kilogram-cal/sec-cm^2	3.184×10^8	BTU/day-ft^2
	1.327×10^7	BTU/hr-ft^2
	3.60×10^6	gram-cal/hr-cm^2
	4186.	watts/cm^2
kilogram-cal-cm/sec-cm^2-°C	6.967×10^7	BTU-in/day-ft^2-°F
	2.903×10^6	BTU-in/hr-ft^2-°F
	4186.	watts-cm/cm^2-°C
kilogram-cm^2	1000.	gram-cm^2
	0.002373	pound-ft^2
	0.3417	pound-in^2
kilogram (force)-m	0.009288	BTU
	96.78	cubic centimeter-atm
	0.003418	cubic foot-atm
	9.80665×10^7	dyne-cm
	9.80665×10^7	ergs
	7.2330	foot-lbf
	232.72	foot-poundals
	2.3427	gram-cal
	100,000.	gram-cm
	3.653×10^{-6}	horsepower-hr (U.S.)
	9.80665	joules
	0.0023423	kilogram-cal
	2.7241×10^{-6}	kilowatt-hr
	0.096784	liter-atm
	98.0665	megalergs
	9.80665	newton-m
	86.799	pound-in
	0.0027241	watt-hr

PROFESSIONAL PUBLICATIONS, INC. ● Belmont, CA

MULTIPLY	BY	TO OBTAIN
kilogram (force)-m/sec	0.55727	BTU/min
	0.009288	BTU/sec
	9.80665×10^7	dyne-cm/sec
	9.80665×10^7	ergs/sec
	26,036.	foot-lbf/hr
	433.98	foot-lbf/min
	7.2330	foot-lbf/sec
	100,000.	gram-cm/sec
	0.013333	horsepower (metric)
	0.013151	horsepower (U.S.)
	9.80665	joules/sec
	0.1405	kilogram-cal/min
	0.0023423	kilogram-cal/sec
	0.0098067	kilowatts
	6555.1	lumens
	9.80665	watts
kilogram (force)-sec/m^2	9806.65	centipoises
	98.0665	dyne-sec/cm^2
	98.0665	grams/cm-sec
	35,304.	kilograms/m-hr
	98.0665	poises
	23,723.	pounds (mass)/ft-hr
	6.590	pounds (mass)/ft-sec
	0.2048	pound-sec/ft^2
	0.001422	pound-sec/in^2
	23,733.	poundal-hr/ft^2
	6.590	poundal-sec/ft^2
kilolines	1000.	lines
	1000.	maxwells
	0.001	megalines
	1.0×10^{-5}	volt-sec
	1.0×10^{-5}	webers
kiloliters	8.3864	barrels (31.5 U.S. gallons)
	35.315	cubic feet
	61,024.	cubic inches
	1.	cubic meters
	1.308	cubic yards
	264.17	gallons (U.S., liquid)
	4.19321	hogsheads
	1000.	liters
	1056.7	quarts (U.S., liquid)

PROFESSIONAL PUBLICATIONS, INC. • Belmont, CA

MULTIPLY	BY	TO OBTAIN
kilometers	1.0×10^{13}	angstroms
	6.684×10^{-9}	astronomical units
	1.0×10^{5}	centimeters
	49.710	chains (Gunter's or surveyor's)
	32.808	chains (Ramsden's or engineer's)
	546.81	fathoms
	3280.84	feet
	4.9710	furlongs
	3.9370×10^{4}	inches
	0.17987	leagues (nautical)
	0.20712	leagues (statute)
	1.0570×10^{-13}	light years
	1000.	meters
	10×10^{9}	micrometers
	0.53961	miles (nautical)
	0.62137	miles (statute)
	1.0×10^{6}	millimeters
	1.0×10^{12}	nanometers
	1312.3	paces
	3.2408×10^{-14}	parsecs
	198.84	perches
	198.84	rods
	9.113	skeins
	1093.61	yards
kilometers/hr	27.778	centimeters/sec
	3280.8	feet/hr
	54.68	feet/min
	0.9113	feet/sec
	0.016667	kilometers/min
	0.5396	knots
	16.667	meters/min
	0.27778	meters/sec
	0.6213712	miles/hr
	0.010356	miles/min
	9.2657×10^{-10}	velocity of light
kilometers/hr-sec	27.7778	centimeters/sec^2
	0.911344	feet/sec^2
	0.0283255	g's (gravities)
	10.936	inches/sec^2
	0.277778	meters/sec^2
	37.2823	miles/hr/min
	0.6214	miles/hr/sec

MULTIPLY	BY	TO OBTAIN
kilometers/min	1666.67	centimeters/sec
	196,850.	feet/hr
	3280.8	feet/min
	54.681	feet/sec
	60.	kilometers/hr
	32.377	knots
	1000.	meters/min
	16.667	meters/sec
	37.282	miles/hr
	0.62137	miles/min
	5.5594×10^{-8}	velocity of light
kilonewtons	0.2248	kips
kilonewtons/m	0.06852	kips/ft
kilowatts	3412.1	BTU/hr
	56.869	BTU/min
	0.94782	BTU/sec
	1.0×10^{10}	dyne-cm/sec
	1.0×10^{10}	ergs/sec
	2.6549×10^{6}	foot-lbf/hr
	44,254.	foot-lbf/min
	737.56	foot-lbf/sec
	1.0197×10^{7}	gram-cm/sec
	1.3596	horsepower (metric)
	1.341	horsepower (U.S.)
	1000.	joules/sec
	14.333	kilogram-cal/min
	0.23885	kilogram-cal/sec
	101.97	kilogram-m/sec
	668,450.	lumens
	1000.	watts

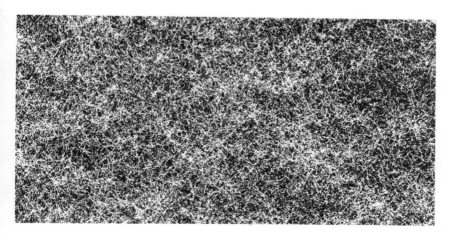

MULTIPLY	BY	TO OBTAIN
kilowatt-hr	3412.	BTU
	3.553×10^7	cubic centimeter-atm
	1254.7	cubic foot-atm
	3.6×10^{13}	dyne-cm
	3.6×10^{13}	ergs
	2.655×10^6	foot-lbf
	8.543×10^7	foot-poundals
	860,000.	gram-cal
	3.671×10^{10}	gram-cm
	1.341	horsepower-hr (U.S.)
	3.600×10^6	joules
	859.8	kilogram-cal
	367,098.	kilogram-m
	35,529.	liter-atm
	3.60×10^7	megalergs
	1000.	watt-hr
kips	453.592	kilograms (force)
	4.4482	kilonewtons
	4448.2	newtons
	1000.	pounds
	32,174.	poundals
	4.4482	sthènes
kips/ft	14.59	kilonewtons/m
kips/ft^2	47.88	kilonewtons/m^2
kips/in^2	6.895	MPa
	1000.	pounds/in^2
	6.895×10^6	pascals
knots (international)	51.44	centimeters/sec
	6070.11	feet/hr
	101.27	feet/min
	1.688	feet/sec
	1.852	kilometers/hr
	0.030867	kilometers/min
	30.867	meters/min
	0.51444	meters/sec
	1.0	miles (nautical)/hr
	1.1508	miles (statute)/hr
	0.019180	miles (statute)/min
ksi	(see kips/in^2)	

PROFESSIONAL PUBLICATIONS, INC. ● Belmont, CA

MULTIPLY	BY	TO OBTAIN
labors	177.14	acres
lamberts	10,000.	apostilbs
	0.31831	candles/cm^2
	295.720	candles/ft^2
	2.05361	candles/in^2
	3183.1	candles/m^2
	929.020	foot-lamberts
	0.31831	lumens/cm^2-sterad
	295.72	lumens/ft^2-sterad
	1000.	millilamberts
	1.0×10^7	nox
	0.31831	stilbs
langleys	41,840.	joules/m^2
leagues (nautical)	3040.	fathoms
	18,240.	feet
	5.5596	kilometers
	1.1515	leagues (statute)
	5559.6	meters
	3.	miles (nautical)
	3.4545	miles (statute)
	6080.2	yards
leagues (statute)	15,840.	feet
	24.	furlongs
	4.8280	kilometers
	0.868421	leagues (nautical)
	4828.0	meters
	2.60526	miles (nautical)
	3.	miles (statute)
	960.	perches
	960.	rods
	5280.	yards
light years	63,242.	astronomical units
	9.4609×10^{12}	kilometers
	5.8787×10^{12}	miles (statute)
	0.30661	parsecs
lines	0.25	barleycorns
	0.211667	centimeters
	12.	douziemes
	0.0069444	feet
	0.083333	inches
	0.001	kilolines
	1.	maxwells
	1.0×10^{-6}	megalines
	1.0×10^{-8}	volt-sec
	1.0×10^{-8}	webers

MULTIPLY	BY	TO OBTAIN
lines/cm^2	0.7958	ampere-turns/cm
	2.0213	ampere-turns/in
	100,000.	gammas
	1.	gauss
	1.	gilberts/cm
	6.452	lines/in^2
	1.0×10^{-4}	teslas
lines/in^2	0.12334	ampere-turns/cm
	0.31329	ampere-turns/in
	15,500.	gammas
	0.155	gauss
	0.155	gilberts/cm
	0.155	lines/cm^2
	0.1550	oersteds
	1.550×10^{-5}	teslas
links (Gunter's or surveyor's)	20.117	centimeters
	0.01	chains (Gunter's or surveyor's)
	0.0066	chains (Ramsden's or engineer's)
	0.66	feet
	7.92	inches
	0.66	links (Ramsden's or engineer's)
	0.201168	meters
	0.264	paces
	0.04	perches
	0.04	rods
	0.22	yards
links (Ramsden's or engineer's)	30.480	centimeters
	0.0151515	chains (Gunter's or surveyor's)
	0.01	chains (Ramsden's or engineer's)
	1.	feet
	12.	inches
	1.51515	links (Gunter's or surveyor's)
	0.30480	meters
	0.4	paces
	0.060606	perches
	0.060606	rods
	0.333333	yards

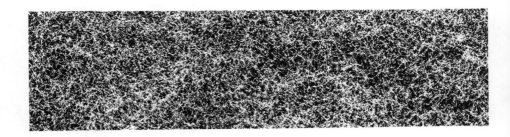

PROFESSIONAL PUBLICATIONS, INC. • Belmont, CA

MULTIPLY	BY	TO OBTAIN
liters	0.00838642	barrels (31.5 U.S. gallons)
	100.	centiliters
	1000.	cubic centimeters
	0.035315	cubic feet
	61.024	cubic inches
	0.001	cubic meters
	1.00×10^6	cubic millimeters
	0.001308	cubic yards
	0.1	decaliters
	10.	deciliters
	270.51	drams
	0.029353	firkins
	0.2270	gallons (U.S., dry)
	0.26417	gallons (U.S., liquid)
	8.4535	gills
	0.0041932	hogsheads
	0.001	kiloliters
	1.0×10^6	microliters
	1000.	millimeters
	16,231.	minims
	33.814	ounces (U.S., liquid)
	1.8162	pints (U.S., dry)
	2.1134	pints (U.S., liquid)
	0.9081	quarts (U.S., dry)
	1.0567	quarts (U.S., liquid)
liters/cm-day	1000.	square centimeters/day
	0.011574	square centimeters/sec
	0.00179397	square inches/sec
liters/min	16.6667	cubic centimeters/sec
	0.03532	cubic feet/min
	5.8858×10^{-4}	cubic feet/sec
	0.001308	cubic yards/min
	0.26417	gallons (U.S., liquid)/min
	0.004403	gallons (U.S., liquid)/sec
	0.016667	liters/sec
	0.02943	miner's inches*
	2.2046	pounds H_2O/min
liters/sec	1000.	cubic centimeters/sec
	2.1189	cubic feet/min
	0.0353147	cubic feet/sec
	0.078478	cubic yards/min
	951.02	gallons (U.S., liquid)/hr
	15.85	gallons (U.S., liquid)/min
	0.26417	gallons (U.S., liquid)/sec
	60.	liters/min
	1.766	miner's inches*
	132.3	pounds H_2O/min

* ID, KS, ND, NE, NM, NV, SD, UT

PROFESSIONAL PUBLICATIONS, INC. ● Belmont, CA

MULTIPLY	BY	TO OBTAIN
liter-atm	0.09607	BTU
	1000.	cubic centimeter-atm
	0.03532	cubic foot-atm
	1.0133×10^9	dyne-cm
	1.0133×10^9	ergs
	74.734	foot-lbf
	2404.5	foot-poundals
	24.206	gram-cal
	1.0332×10^6	gram-cm
	3.774×10^{-5}	horsepower-hr (U.S.)
	101.325	joules
	0.02420	kilogram-cal
	10.332	kilogram-m
	2.8146×10^{-5}	kilowatt-hr
	1013.3	megalergs
	0.02815	watt-hr
lumens	8.5116×10^{-5}	BTU/min
	1.4186×10^{-6}	BTU/sec
	14,960.	dyne-cm/sec
	14,960.	ergs/sec
	3.9722	foot-lbf/hr
	0.066204	foot-lbf/min
	0.00110338	foot-lbf/sec
	15.255	gram-cm/sec
	2.0341×10^{-6}	horsepower (metric)
	2.0061×10^{-6}	horsepower (U.S.)
	0.001496	joules/sec
	2.1443×10^{-5}	kilogram-cal/min
	3.574×10^{-7}	kilogram-cal/sec
	1.526×10^{-4}	kilogram-m/sec
	1.496×10^{-6}	kilowatts
	0.001470	watts
lumens/cm^2	929.02	foot-candles
	929.02	lumens/ft^2
	6.4516	lumens/in^2
	10,000.	lumens/m^2
	10,000.	lux
	10,000.	meter-candles
	1000.	milliphots
	1.	phots

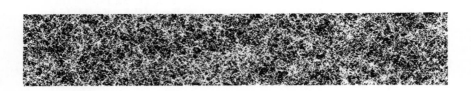

PROFESSIONAL PUBLICATIONS, INC. • Belmont, CA

MULTIPLY	BY	TO OBTAIN
lumens/cm^2-sterad	31,415.9	apostilbs
	1.	candles/cm^2
	929.03	candles/ft^2
	6.45157	candles/in^2
	10,000.	candles/m^2
	2918.6	foot-lamberts
	3.14159	lamberts
	929.026	lumens/ft^2-sterad
	3141.59	millilamberts
	1.	stilbs
lumens/ft^2	1.	foot-candles
	0.0010764	lumens/cm^2
	0.0069444	lumens/in^2
	10.764	lumens/m^2
	10.764	lux
	10.764	meter-candles
	1.0764	milliphots
	0.0010764	phots
lumens/ft^2-sterad	33.816	apostilbs
	0.0010764	candles/cm^2
	1.	candles/ft^2
	0.0069444	candles/in^2
	10.764	candles/m^2
	3.14159	foot-lamberts
	0.0033816	lamberts
	0.0010764	lumens/cm^2-sterad
	3.3816	millilamberts
	0.0010764	stilbs
lumens/in^2	144.	foot-candles
	0.1550	lumens/cm^2
	144.	lumens/ft^2
	1550.	lumens/m^2
	1550.	lux
	155.	milliphots
	0.1550	phots

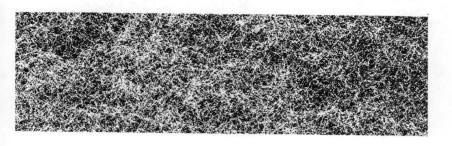

PROFESSIONAL PUBLICATIONS, INC. • Belmont, CA

MULTIPLY	BY	TO OBTAIN
lumens/m^2	0.092902	foot-candles
	1.0×10^{-4}	lumens/cm^2
	0.092902	lumens/ft^2
	6.4516×10^{-4}	lumens/in^2
	1.	lux
	1.	meter-candles
	0.1	milliphots
	1.0×10^{-4}	phots
lumens/sterad	1.	bougie decimales
	1.	candles (int'l)
	0.104	carcel units
	1.	English sperm candles
	1.11111	hefner units
	1.	pentane candles
lux	0.092902	foot-candles
	1.0×10^{-4}	lumens/cm^2
	0.092902	lumens/ft^2
	6.4516×10^{-4}	lumens/in^2
	1.	lumens/m^2
	1.	meter-candles
	0.1	milliphots
	1000.	nox
	1.0×10^{-4}	phots

MULTIPLY	BY	TO OBTAIN
maxwells	0.001	kilolines
	1.	lines
	1.0×10^{-6}	megalines
	1.0×10^{-8}	volt-sec
	1.0×10^{-8}	webers
maxwells/cm^2	1.	gauss
maxwells/in^2	0.1550	gauss
megalergs	9.481×10^{-5}	BTU
	0.9869	cubic centimeter-atm
	3.485×10^{-5}	cubic foot-atm
	1.0×10^{6}	dyne-cm
	1.0×10^{6}	ergs
	0.07376	foot-lbf
	2.373	foot-poundals
	0.02389	gram-cal
	1020.	gram-cm
	3.725×10^{-8}	horsepower-hr (U.S.)
	0.1	joules
	2.389×10^{-5}	kilogram-cal
	0.01020	kilogram-m
	2.7778×10^{-8}	kilowatt-hr
	9.869×10^{-4}	liter-atm
	2.7778×10^{-5}	watt-hr
megalines	1000.	kilolines
	1.0×10^{6}	lines
	1.0×10^{6}	maxwells
	0.01	volt-sec
	0.01	webers
megohms (megs)	1.0×10^{15}	abohms
	1.0×10^{12}	microhms
	1.0×10^{6}	ohms
	1.1126×10^{-6}	statohms
Meinzer units	1.00	gallons/day-ft^2
mercury at 0°C	5.94103×10^{6}	grains/ft^3
	13.5951	grams/cm^3
	1.35951×10^{7}	grams/m^3
	13.5951	grams/ml
	13,595.1	kilograms/m^3
	848.738	pounds/ft^3
	113.453	pounds/gallon (U.S., liquid)
	0.491188	pounds/in^3
	4.62911×10^{-6}	pounds/mil foot
	26.3776	slugs/ft^3

MULTIPLY	BY	TO OBTAIN
meters	1.0×10^{10}	angstroms
	6.684×10^{-12}	astronomical units
	0.027340	bolts
	0.0045567	cable lengths
	100.	centimeters
	0.04971	chains (Gunther's or surveyor's)
	0.03281	chains (Ramsden's or engineer's)
	2.1872	cubits
	0.1	decameters
	10.	decimeters
	0.87489	ells
	0.54681	fathoms
	3.28084	feet
	1.0×10^{15}	fermis
	0.004971	furlongs
	9.8425	hands
	39.3701	inches
	0.001	kilometers
	4.9710	links (Gunter's or surveyor's)
	3.2808	links (Ramsden's or engineer's)
	1.0×10^{6}	micrometers
	1.0×10^{12}	micromicrons
	1.0×10^{6}	microns
	5.3961×10^{-4}	miles (nautical)
	6.2137×10^{-4}	miles (statute)
	1.0×10^{9}	millimicrons
	1000.	millimeters
	39,370.1	mils
	1.0×10^{9}	nanometers
	1.31	paces
	3.2408×10^{-17}	parsecs
	0.198838	perches
	0.198838	rods
	0.16404	ropes
	0.0091134	skeins
	1.0×10^{12}	stigmas
	1.0×10^{10}	tenthmeters
	1.18110	varas
	1.553×10^{6}	wavelengths red line cadmium
	9.980×10^{12}	X-units
	1.0936	yards
	1.0×10^{15}	yukawas

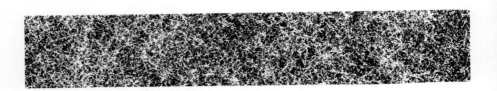

PROFESSIONAL PUBLICATIONS, INC. • Belmont, CA

MULTIPLY	BY	TO OBTAIN
meters/min	1.6667	centimeters/sec
	196.85	feet/hr
	3.281	feet/min
	0.05468	feet/sec
	0.06	kilometers/hr
	0.001	kilometers/min
	0.032375	knots
	0.016667	meters/sec
	0.03728	miles/hr
	6.2137×10^{-4}	miles/min
meters/sec	100.	centimeters/sec
	11,811.	feet/hr
	196.9	feet/min
	3.281	feet/sec
	3.6	kilometers/hr
	0.06	kilometers/min
	1.9425	knots
	60.	meters/min
	2.2369	miles/hr
	0.03728	miles/min
meters/sec^2	100.	centimeters/sec^2
	3.2808	feet/sec^2
	0.101972	g (gravity)
	39.37	inches/sec^2
	3.6	kilometers/hr-sec
	134.216	miles/hr-min
	2.237	miles/hr-sec
meter-candles	0.092902	foot-candles
	1.0×10^{-4}	lumens/cm^2
	0.092902	lumens/ft^2
	1.	lumens/m^2
	1.	lux
	0.1	milliphots
	1.0×10^{-4}	phots
meter-kilograms (force)	9.8067	joules
MeV (million electron volts)	1.0×10^6	electron volts
MGD (million gallons per day)	1.0×10^6	gallons/day
	1.5472	cubic feet/sec
	0.04381	cubic meters/sec
MGD/acre	22.96	gallons/day-ft^2
mhos	1.0×10^{-9}	abmhos
	1.0×10^6	megmhos
	1.0×10^{-6}	micromhos
	1.	siemens

PROFESSIONAL PUBLICATIONS, INC. ● Belmont, CA

MULTIPLY	BY	TO OBTAIN
microfarads	1.0×10^{-15}	abfarads
	1.0×10^{-6}	farads
	1.0×10^{6}	micromicrofarads
	1.0×10^{6}	picofarads
	898,755.	statfarads
micrograms	6.02×10^{17}	daltons
	1.0×10^{-6}	grams
	1.0×10^{-9}	kilograms
	0.001	milligrams
micrograms/ml	16.691	drams/ounce (U.S., liquid)
	456.39	grains/ounce (U.S., liquid)
	0.001	grams/ℓ
	1.	grams/m^3
	0.001	kilograms/m^3
	1.	milligrams/ℓ
	33.382	ounces/quart (U.S., liquid)
	6.2428×10^{-5}	pounds/ft^3
	3.613×10^{-8}	pounds/in^3
microhenrys	1.0×10^{3}	abhenrys
	1.0×10^{-6}	henrys (absolute)
	1.0×10^{-3}	millihenrys
	1.11265×10^{-18}	stathenrys
microhms	1000.	abohms
	1.0×10^{-12}	megohms
	1.0×10^{-6}	ohms
	1.1127×10^{-18}	statohms
microliters	6.1024×10^{-5}	cubic inches
	1.	cubic millimeters
	2.7052×10^{-4}	drams
	1.0×10^{-6}	liters
	0.001	milliliters
	0.016231	minims
	3.3814×10^{-5}	ounces (U.S., liquid)

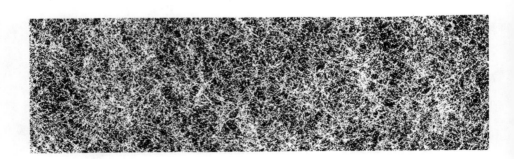

MULTIPLY	BY	TO OBTAIN
micrometers	10,000.	angstroms
	1.0×10^{-4}	centimeters
	3.2808×10^{-6}	feet
	3.9370×10^{-5}	inches
	1.0×10^{-9}	kilometers
	1.0×10^{-6}	meters
	1.0×10^{6}	micromicrons
	6.214×10^{-7}	miles
	0.001	millimeters
	1000.	millimicrons
	0.0393701	mils
	1000.	nanometers
	1.0×10^{6}	stigmas
micromicrofarads	1.0×10^{-21}	abfarads
	1.0×10^{-12}	farads
	1.0×10^{-6}	microfarads
	1.	picofarads
	0.898776	statfarads
micromicrons	0.01	angstroms
	3.28084×10^{-12}	feet
	3.93701×10^{-11}	inches
	1.0×10^{-12}	meters
	1.0×10^{-6}	microns
	1.0×10^{-9}	millimeters
	0.001	millimicrons
	3.93701×10^{-8}	mils
	1.	stigmas
microns	(see micrometers)	
mils	1.0×10^{-3}	inches
	2.54×10^{-5}	meters
	25.4	micrometers
miles	5.2188×10^{-14}	parsecs
miles (U.S. nautical)	8.4444	cable lengths
	1013.3	fathoms
	6080.2	feet
	1.85318	kilometers
	0.33333	leagues (nautical)
	0.38384	leagues (statute)
	1853.18	meters
	1.1515	miles (U.S. statute)
	2026.7	yards

PROFESSIONAL PUBLICATIONS, INC. ● Belmont, CA

MULTIPLY	BY	TO OBTAIN
miles (statute)	1.609×10^{13}	angstroms
	1.076×10^{-8}	astronomical units
	1.609×10^{5}	centimeters
	80.	chains (Gunter's or surveyor's)
	52.8	chains (Ramsden's or engineer's)
	880.	fathoms
	5280.	feet
	8.	furlongs
	63,360.	inches
	1.60934	kilometers
	0.289474	leagues (nautical)
	0.333333	leagues (statute)
	1.7011×10^{-13}	light years
	8000.	links (Gunter's or surveyor's)
	5280.	links (Ramsden's or engineer's)
	1609.34	meters
	1.609×10^{9}	micrometers
	0.86842	miles (nautical)
	1.609×10^{6}	millimeters
	1.609×10^{12}	nanometers
	2112.	paces
	320.	perches
	320.	rods
	14.6667	skeins
	1760.	yards
miles/hr	44.704	centimeters/sec
	5280.	feet/hr
	88.	feet/min
	1.46667	feet/sec
	1.6093	kilometers/hr
	0.026822	kilometers/min
	0.8684	knots
	26.82	meters/min
	0.44704	meters/sec
	0.016667	miles/min
	1.4912×10^{-9}	velocity of light
miles/hr-min	0.74507	centimeters/sec^2
	0.0244444	feet/sec^2
	7.5976×10^{-4}	g (gravity)
	0.29333	inches/sec^2
	0.0268224	kilometers/hr-sec
	0.0074507	meters/sec^2
	0.0166667	miles/hr-sec

MULTIPLY	BY	TO OBTAIN
miles/hr-sec	44.704	centimeters/sec^2
	1.46667	feet/sec^2
	0.0455854	g (gravity)
	17.6	inches/sec^2
	1.60934	kilometers/hr-sec
	0.44704	meters/sec^2
	60.	miles/hr-min
miles/min	2682.2	centimeters/sec
	316,800.	feet/hr
	5280.	feet/min
	88.	feet/sec
	96.561	kilometers/hr
	1.609	kilometers/min
	52.105	knots
	1609.	meters/min
	26.822	meters/sec
	60.	miles/hr
	8.9470×10^{-8}	velocity of light
millennia	10.	centuries
	100.	decades
	1000.	years
milliamperes	1.0×10^{-4}	abamperes
	0.001	amperes
	1.0363×10^{-8}	faradays/sec
	2.99796×10^6	statamperes
milligrams	0.001	grams
	1.0×10^{-6}	kilograms
	1000.	micrograms
	2.205×10^{-6}	pounds
milligrams/g	1016.05	grams/ton (long)
	1000.	grams/ton (metric)
	977.52	grams/ton (short)
	0.214286	karats
	1000.	milligrams/kg
	29.175	milligrams/ton (assay)
	35.840	ounces/ton (long)
	32.	ounces/ton
	2.24	pounds/ton (long)
	2.	pounds/ton (short)

PROFESSIONAL PUBLICATIONS, INC. • Belmont, CA

MULTIPLY	BY	TO OBTAIN
milligrams/kg	1.01605	grams/ton (long)
	1.	grams/ton (metric)
	0.977517	grams/ton (short)
	2.14286×10^{-4}	karats
	0.001	milligrams/g
	0.029175	milligrams/ton (assay)
	0.035840	ounces/ton (long)
	0.032	ounces/ton (short)
	0.002240	pounds/ton (long)
	0.002	pounds/ton (short)
milligrams/ℓ (mg/ℓ)	0.07	degrees, Clarke
	16.691	drams/ounce (U.S., liquid)
	0.0584	grains/gallon
	456.39	grains/ounce (U.S., liquid)
	0.001	grams/ℓ
	1.	grams/m^3
	0.001	kilograms/m^3
	1.	micrograms/ml
	33.382	ounces/quart (U.S., liquid)
	1.	parts per million (ppm)
	6.2428×10^{-5}	pounds/ft^3
	3.613×10^{-8}	pounds/in^3
	8.345	pounds/million gallons
milligrams/ton (assay)	34.826	grams/ton (long)
	34.276	grams/ton (metric)
	33.505	grams/ton (short)
	0.00734485	karats
	0.034276	milligrams/g
	34.276	milligrams/kg
	1.22845	ounces/ton (long)
	1.0968	ounces/ton (short)
	0.076778	pounds/ton (long)
	0.068552	pounds/ton (short)
milligram weights/in	0.38609	dynes/cm
	0.38609	ergs/cm^2
	0.0038609	ergs/mm^2
	0.03937	milligram wts/mm

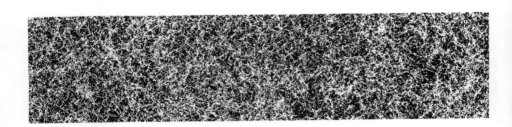

PROFESSIONAL PUBLICATIONS, INC. • Belmont, CA

MULTIPLY	BY	TO OBTAIN
milligram weights/mm	9.80665	dynes/cm
	9.80665	ergs/cm^2
	0.0980665	ergs/mm^2
	25.4	milligram wts/in
millihenrys	1.0×10^6	abhenrys
	0.001	henrys (absolute)
	1000.	microhenrys
	1.11265×10^{-15}	stathenrys
millilamberts	3.1831×10^{-4}	candles/cm^2
	0.0020536	candles/in^2
	0.001	lamberts
	3.1831×10^{-4}	lumens/cm^2-sterad
	0.29572	lumens/ft^2-sterad
	3.1831×10^{-4}	stilbs
milliliters	1.	cubic centimeters
	0.061024	cubic inches
	0.27051	drams
	2.6417×10^{-4}	gallons (U.S., liquid)
	0.001	liters
	1000.	microliters
	16.231	minims
	0.033814	ounces (U.S., liquid)
	0.0021134	pints (U.S., liquid)
	0.0010567	quarts (U.S., liquid)
millimeters	1.0×10^7	angstroms
	0.1	centimeters
	0.23711	ems (pica)
	0.0032808	feet
	0.0084538	gills
	0.039370	inches
	1.0×10^{-6}	kilometers
	0.001	meters
	1000.	micrometers
	1.0×10^9	micromicrons
	1000.	microns
	6.214×10^{-7}	miles
	1.0×10^6	millimicrons
	39.3701	mils
	1.0×10^6	nanometers
	2.8453	points
millimeters/meter	0.012	inches/ft

PROFESSIONAL PUBLICATIONS, INC. • Belmont, CA

MULTIPLY	BY	TO OBTAIN
millimeters of mercury	9.678	atmospheres
	0.0032809	feet of water
	0.0393701	inches of mercury
	9.806365	pascals
	0.2048102	pounds per square foot
	0.001422243	pounds per square inch
	0.07355377	torr
millimicrons	10.	angstroms
	3.2808×10^{-9}	feet
	3.9370×10^{-8}	inches
	1.0×10^{-9}	meters
	1000.	micromicrons
	0.001	microns
	1.0×10^{-6}	millimeters
	3.9370×10^{-5}	mils
millions of gallons per day	(see MGD)	
millionth microns	1.	stigmas
milliphots	0.92902	foot-candles
	0.001	lumens/cm^2
	0.92902	lumens/ft^2
	0.0064516	lumens/in^2
	10.	lumens/m^2
	10.	lux
	10.	meter-candles
	0.001	phots
millivolts	1.0×10^5	abvolts
	1000.	microvolts
	3.336×10^{-6}	statvolts
	0.001	volts
mils	254,000.	angstroms
	0.05625	degrees of arc
	0.006	ems (pica)
	8.33333×10^{-5}	feet
	0.001	inches
	2.54×10^{-5}	meters
	2.54×10^7	micromicrons
	25.400	microns
	0.0254	millimeters
	25,400.	millimicrons
	3.038×10^6	minutes
	0.072	points

MULTIPLY	BY	TO OBTAIN
miner's inches (AZ, CA, MT, OR)	1.5	cubic feet/min
	0.025	cubic feet/sec
	11.221	gallons (U.S., liquid)/min
	42.475	liters/min
	0.7079	liters/sec
miner's inches (CO)	1.5624	cubic feet/min
	0.02604	cubic feet/sec
	11.687	gallons (U.S., liquid)/min
	44.239	liters/min
	0.7373	liters/sec
miner's inches (ID, KS, ND, NE, NM, NV, SD, UT)	1.2	cubic feet/min
	0.02	cubic feet/sec
	8.9767	gallons (U.S., liquid)/min
	33.980	liters/min
	0.56632	liters/sec
minims	0.061612	cubic centimeters
	0.0037598	cubic inches
	61.612	cubic millimeters
	0.016667	drams
	1.6276×10^{-5}	gallons (U.S., liquid)
	5.2083×10^{-4}	gills
	6.1612×10^{-5}	liters
	61.612	microliters
	0.0616115	milliliters
	0.0020833	ounces (U.S., liquid)
	1.3021×10^{-4}	pints (U.S., liquid)
	6.5104×10^{-5}	quarts (U.S., liquid)
minutes (angular)	0.0166667	degrees
	0.185185	grades
	0.29630	mils
	1.8519×10^{-4}	quadrants
	2.90888×10^{-4}	radians
	4.62963×10^{-5}	revolutions
	60.	seconds
	5.5556×10^{-4}	signs

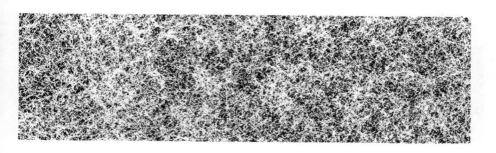

PROFESSIONAL PUBLICATIONS, INC. • Belmont, CA

MULTIPLY	BY	TO OBTAIN
minutes (solar)	6.9635×10^{-4}	days (sidereal)
	6.9444×10^{-4}	days (solar)
	0.016667	hours (solar)
	2.3516×10^{-5}	months (lunar)
	2.2831×10^{-5}	months (mean)
	60.164	seconds (sidereal)
	60.	seconds (solar)
	9.9206×10^{-5}	weeks
	1.8974×10^{-6}	years (leap)
	1.9013×10^{-6}	years (sidereal)
	1.90133×10^{-6}	years (solar)
minutes/cm	0.016667	degrees/cm
	0.50802	degrees/ft
	0.042325	degrees/in
	2.9089×10^{-4}	radians/cm
months (lunar)	29.612	days (sidereal)
	29.530	days (solar)
	708.73	hours (solar)
	42,524.	minutes/solar
	0.970867	months (mean)
	2.5584×10^{6}	seconds (sidereal)
	2.5514×10^{6}	seconds (solar)
	4.2186	weeks
	0.080685	years (leap)
	0.080849	years (sidereal)
	0.080852	years (solar)
months (mean)	30.500	days (sidereal)
	30.42	days (solar)
	730.	hours (solar)
	43,800.	minutes (solar)
	1.0300	months (lunar)
	2.6352×10^{6}	seconds (sidereal)
	2.628×10^{6}	seconds (solar)
	4.3453	weeks
	0.083108	years (leap)
	0.083275	years (sidereal)
	0.083278	years (solar)

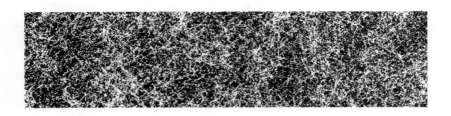

PROFESSIONAL PUBLICATIONS, INC. ● Belmont, CA

MULTIPLY	BY	TO OBTAIN
nanometers	10.	angstroms
	1.0×10^{-7}	centimeters
	3.28084×10^{-9}	feet
	3.937×10^{-8}	inches
	1.0×10^{-12}	kilometers
	1.0×10^{-9}	meters
	0.001	micrometers
	0.001	microns
	6.214×10^{-13}	miles
	1.0×10^{-6}	millimeters
newtons	10.	crinals
	100,000.	dynes
	1573.7	grains
	101.97	grams
	0.10197	kilograms (force)
	1.	kilograms-m/s^2
	2.248×10^{-4}	kips
	3.597	ounces (force)
	0.22481	pounds
	7.233	poundals
	0.001	sthènes
	1.0036×10^{-4}	tons (long)
	1.1241×10^{-4}	tons (short)
newtons/m	1000.	dynes/cm
	0.068522	pounds/ft
	5.71017×10^{-3}	pounds/in
newtons/m^2	9.869×10^{-6}	atmospheres
	1.0×10^{-5}	bars
	10.	baryes
	7.50×10^{-4}	centimeters Hg at 0°C
	10.	dynes/cm^2
	2.95×10^{-4}	inches Hg at 32°F
	0.10197	kilograms (force)/m^2
	1.	pascals
	0.02089	pounds/ft^2
	1.45×10^{-4}	pounds/in^2
	1.044×10^{-5}	tons (short)/ft^2
newton-meters	(see joules)	

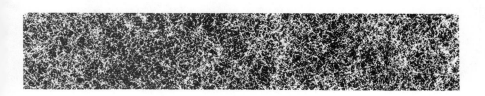

PROFESSIONAL PUBLICATIONS, INC. • Belmont, CA

MULTIPLY	BY	TO OBTAIN
nits	3.14159	apostilbs
	0.092903	candles/ft^2
	6.4516×10^{-4}	candles/in^2
	1.	candles/m^2
	0.29186	foot-lamberts
	3.14159×10^{-4}	lamberts
	1.0×10^{-4}	lumens/cm^2-sterad
	0.092903	lumens/ft^2-sterad
	1.0×10^{-4}	stilbs
nox	9.290×10^{-5}	foot-candle
	1.0×10^{-7}	lamberts
	0.001	lux
	1.0×10^{-7}	phots

PROFESSIONAL PUBLICATIONS, INC. ● Belmont, CA

MULTIPLY	BY	TO OBTAIN
oersteds	79.5775	amperes/m
	2.998×10^{10}	statoersteds
ohms	1.0×10^9	abohms
	1.0×10^{-6}	megohms
	1.0×10^6	microhms
	1.1127×10^{-12}	statohms
ohm-sec	1.	henrys
ounces (avoir)	141.75	carats
	0.000625	centals
	2835.	centigrams
	1.70756×10^{25}	daltons
	16.	drams (avoir)
	7.2917	drams (troy)
	437.5	grains
	28.3495	grams
	5.58036×10^{-4}	hundredweights (long)
	6.25×10^{-4}	hundredweights (short)
	0.0283495	kilograms
	28,349.5	milligrams
	0.9114581	ounces (troy)
	18.229	pennyweights
	0.0625	pounds (avoir)
	0.075955	pounds (troy)
	1.1161×10^{-4}	quarters (long)
	1.25×10^{-4}	quarters (short)
	2.8350×10^{-4}	quintals
	21.875	scruples
	0.0019426	slugs
	0.9720	tons (assay)
	2.7902×10^{-5}	tons (long)
	2.8350×10^{-5}	tons (metric)
	3.125×10^{-5}	tons (short)
ounces (British, liquid)	0.96076	ounces (U.S., liquid)
ounces (force)	0.2780	newtons

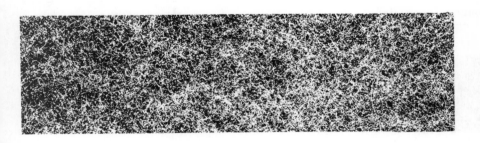

PROFESSIONAL PUBLICATIONS, INC. ● Belmont, CA

MULTIPLY	BY	TO OBTAIN
ounces (troy)	1.0664	assay (troy)
	155.517	carats
	17.554	drams (avoir)
	8.	drams (troy)
	480.	grains
	31.10348	grams
	0.03110348	kilograms
	1.097143	ounces (avoir)
	20.	pennyweights
	0.0685714	pounds (avoir)
	0.083333	pounds (troy)
	24.	scruples
	0.0021313	slugs
	3.0612×10^{-5}	tons (long)
	3.1104×10^{-5}	tons (metric)
	3.4286×10^{-5}	tons (short)
ounces (U.S., liquid)	2.4802×10^{-4}	barrels (31.5 U.S. gallons)
	29.5735	cubic centimeters
	0.0010444	cubic feet
	1.8047	cubic inches
	8.	drams
	8.6806×10^{-4}	firkins
	0.0078125	gallons (U.S., liquid)
	0.25	gills
	0.029574	liters
	29.574	milliliters
	480.	minims
	1.0408	ounces (British, liquid)
	0.0625	pints (U.S., liquid)
	0.03125	quarts (U.S., liquid)
ounces (avoir)/ft^2	0.0069445	ounces (avoir)/in^2
	0.0625	pounds/ft^2
	4.3403×10^{-4}	pounds/in^2
ounces (U.S., liquid)/ft^2	2.954×10^{-5}	atmospheres
	2.993×10^{-5}	bars
	29.93	dynes/cm^2
	0.3052	kilograms/m^2
ounces (avoir)/in^2	0.0042530	atmospheres

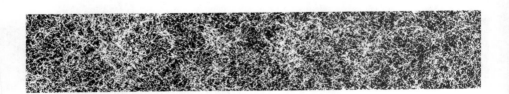

PROFESSIONAL PUBLICATIONS, INC. ● Belmont, CA

MULTIPLY	BY	TO OBTAIN
ounces (U.S., liquid)/in^2	0.004309	bars
	0.32323	centimeters Hg at 0°C
	4309.2	dynes/cm^2
	4.394	grams/cm^2
	0.12726	inches Hg at 32°F
	0.0043942	kilograms/cm^2
	43.942	kilograms/m^2
	4.3942×10^{-5}	kilograms/mm^2
	9.	pounds/ft^2
	0.0625	pounds/in^2
	0.00450	tons (short)/ft^2
	3.125×10^{-5}	tons (short)/in^2
ounces (U.S., liquid)/in^3	28.875	drams/ounce (U.S., liquid)
	789.55	grains/ounce (U.S., liquid)
	1730.	kilograms/m^3
	57.750	ounces/quart (U.S., liquid)
	108.	pounds/ft^3
	0.0625	pounds/in^3
ounces/gallon (U.S., liquid)	3.4180	grains/ounce (U.S., liquid)
	437.5	grains/gallon (U.S., liquid)
	7.4892	kilograms/m^3
	0.25	ounces/quart (U.S., liquid)
	0.46753	pounds/ft^3
	2.7056×10^{-4}	pounds/in^3
ounces (avoir)/gallon (U.S., liquid)	0.125	drams (avoir)/ounce (U.S., liquid)
ounces/quart (U.S., liquid)	0.5	drams/ounce (U.S., liquid)
	13.672	grains/ounce (U.S., liquid)
	29.957	kilograms/m^3
	4.	ounces/gallon (U.S., liquid)
	0.017316	ounces (U.S., liquid)/in^3
	1.87013	pounds/ft^3
	0.0010823	pounds/in^3
ounces (U.S., liquid)/ton (long)	28.350	grams/ton (long)
	27.902	grams/ton (metric)
	27.2745	grams/ton (short)
	0.0059790	karats
	0.027902	milligrams/g
	27.902	milligrams/kg
	0.814035	milligrams/ton (assay)
	0.89286	ounces (U.S., liquid)/ton (short)
	0.0625	pounds/ton (long)
	0.055804	pounds/ton (short)

PROFESSIONAL PUBLICATIONS, INC. • Belmont, CA

MULTIPLY	BY	TO OBTAIN
ounces (U.S., liquid)/ton (short)	31.752	grams/ton (long)
	31.25	grams/ton (metric)
	30.547	grams/ton (short)
	0.0066964	karats
	0.03125	milligrams/g
	31.25	milligrams/kg
	0.911719	milligrams/ton (assay)
	1.12	ounces (U.S., liquid)/ton (long)
	0.07	pounds/ton (long)
	0.0625	pounds/ton (short)
outs	330.	feet
	10.	33-foot chains

PROFESSIONAL PUBLICATIONS, INC. ● Belmont, CA

MULTIPLY	BY	TO OBTAIN
paces	76.2000	centimeters
	0.0379	chains (Gunter's or surveyor's)
	0.025	chains (Ramsden's or engineer's)
	2.5	feet
	0.00379	furlongs
	30.	inches
	7.62×10^{-4}	kilometers
	3.79	links (Gunter's or surveyor's)
	2.5	links (Ramsden's or engineer's)
	0.762	meters
	4.73486×10^{-4}	miles (statute)
	0.15	perches
	0.15	rods
	0.833333	yards
palms	7.62	centimeters
	4.	digits
	0.25	feet
	0.75	hands
	3.	inches
	0.33333	spans
parsecs	206,270.	astronomical units
	3.08568×10^{13}	kilometers
	3.2615	light years
	3.08568×10^{16}	meters
	1.91735×10^{13}	miles
parts per million	(see milligrams/ℓ)	
pascals	9.8692×10^{-6}	atmospheres
	1.0×10^{-5}	bars
	10.	baryes
	10.	dynes/cm^2
	4.022×10^{-3}	inches H_2O
	2.964×10^{-4}	inches Hg at 32°
	0.101972	kilograms (force)/m^2
	1.0×10^{-3}	kilopascals
	1.45×10^{-7}	kips/in^2
	1.0×10^{-6}	megapascals
	1.	newtons/m^2
	0.020885	pounds/ft^2
	1.4504×10^{-4}	pounds/in^2
	0.67197	poundals/ft^2
	1.044×10^{-5}	tons (short)/ft^2
	0.007501	torrs
pascal-sec	1.0×10^3	centipoises

MULTIPLY	BY	TO OBTAIN
pecks	0.25	bushels
	0.31111	cubic feet
	537.605	cubic inches
	0.0088098	cubic meters
	0.011523	cubic yards
	2.	gallons (U.S., dry)
	2.3273	gallons (U.S., liquid)
	16.	pints (U.S., dry)
	8.	quarts (U.S., dry)
	0.0088098	steres
pennyweights	7.77587	carats
	0.877714	drams (avoir)
	0.4	drams (troy)
	24.	grains
	1.55517	grams
	0.00155517	kilograms
	1555.17	milligrams
	0.054857	ounces (avoir)
	0.05	ounces (troy)
	0.0034286	pounds (avoir)
	0.00416667	pounds (troy)
	1.2	scruples
	0.05332	tons (assay)
pentane candles	1.	bougie decimales
	1.	candles (int'l)
	0.104167	carcel units
	1.	English sperm candles
	1.11111	hefner units
	1.	lumens/sterad
perches	16.5	feet
	198.	inches
	5.0292	meters
	1.	poles
	1.	rods
	5.5	yards
phots	929.02	foot-candles
	1.	lumens/cm^2
	929.02	lumens/ft^2
	6.4516	lumens/in^2
	10,000.	lumens/m^2
	10,000.	lux
	10,000.	meter-candles
	1000.	milliphots
	1.0×10^7	nox

MULTIPLY	BY	TO OBTAIN
picofarads	1.0×10^{-21}	abfarads
	1.0×10^{-12}	farads
	1.0×10^{-6}	microfarads
	1.	micromicrofarads
	0.898776	statfarads
pints (U.S., dry)	0.015625	bushels
	550.61	cubic centimeters
	0.019445	cubic feet
	33.6	cubic inches
	5.5060×10^{-4}	cubic meters
	7.2017×10^{-4}	cubic yards
	0.125	gallons (U.S., dry)
	0.14546	gallons (U.S., liquid)
	0.0625	pecks
	0.5	quarts (U.S., dry)
pints (U.S., liquid)	0.0039683	barrels (31.5 U.S. gallons)
	473.18	cubic centimeters
	0.016710	cubic feet
	28.875	cubic inches
	4.7318×10^{-4}	cubic meters
	6.1889×10^{-4}	cubic yards
	2.	cups
	128.	drams
	0.0138889	firkins
	0.10742	gallons (U.S., dry)
	0.125	gallons (U.S., liquid)
	4.	gills
	0.47318	liters
	473.176	milliliters
	7680.	minims
	16.	ounces (U.S., liquid)
	0.5	quarts (U.S., liquid)
	4.7318×10^{-4}	steres
pipes	4.	barrels (31.5 U.S. gallons)
	1.	butts
	126.1	gallons (U.S., liquid)
	2.	hogsheads
points	0.0351460	centimeters
	0.0833333	ems (pica)
	0.013837	inches
	0.351460	millimeters
	13.837	mils

PROFESSIONAL PUBLICATIONS, INC. • Belmont, CA

MULTIPLY	BY	TO OBTAIN
poises	100.	centipoises
	1.	dyne-sec/cm^2
	0.0010197	gram (force)-sec/cm^2
	1.	grams/sec-cm
	0.0101972	kilogram (force)-sec/m^2
	360.	kilograms/m-hr
	0.1	pascal-sec
	241.91	pounds (mass)/ft-hr
	0.067197	pounds (mass)/ft-sec
	0.002089	pound-sec/ft^2
	1.450×10^{-5}	pound-sec/in^2
	241.93	poundal-hr/ft^2
	0.067197	poundal-sec/ft^2
poise-cm^3/g	0.016018	poise-ft^3/lbm
	0.061024	poise-in^3/g
	1.	square centimeters/sec
	0.1550	square inches/sec
poise-ft^3/lbm	62.43	poise-cm^3/g
	3.8096	poise-in^3/g
	62.43	square centimeters/sec `
	9.6776	square inches/sec
poise-in^3/g	16.387	poise-cm^3/g
	0.26250	poise-ft^3/lbm
	16.387	square centimeters/sec
	2.54	square inches/sec
poles	16.5	feet
	5.0292	meters
	1.	perches
	1.	rods
	5.5	yards

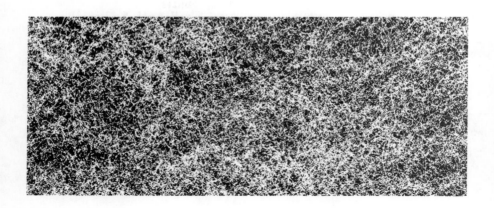

PROFESSIONAL PUBLICATIONS, INC. ● Belmont, CA

MULTIPLY	BY	TO OBTAIN
pounds (mass, avoir)	2267.96	carats
	0.01	centals
	2.732×10^{26}	daltons
	256.	drams (avoir)
	116.667	drams (troy)
	7000.	grains
	453.59237	grams
	0.0089286	hundredweights (long)
	0.01	hundredweights (short)
	0.45359237	kilograms
	0.001	kips
	4.536×10^5	milligrams
	16.	ounces (avoir)
	14.5833	ounces (troy)
	291.667	pennyweights
	1.21528	pounds (troy)
	0.0017857	quarters (long)
	0.002	quarters (short)
	0.00453592	quintals
	350.	scruples
	0.031081	slugs
	0.07143	stones
	4.4643×10^{-4}	tons (long)
	4.5359×10^{-4}	tons (metric)
	5.0×10^{-4}	tons (short)
pounds (force)	44.482	crinals
	444,822.	dynes
	0.001	kips
	4.4482	newtons
	32.174	poundals
	0.004448	sthènes
pounds (force)/ft	14,593.8	dynes/cm
	14.59383	newtons/m
	12.0	pounds/ft
pounds (force)/in	175,126.	dynes/cm
	175.126	newtons/m
	12.0	pounds/ft

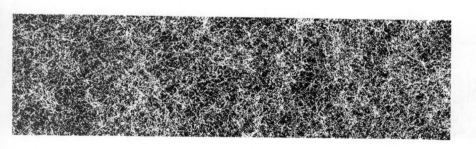

PROFESSIONAL PUBLICATIONS, INC. ● Belmont, CA

MULTIPLY	BY	TO OBTAIN
pounds (mass, troy)	1866.2	carats
	210.652	drams (avoir)
	96.	drams (troy)
	5760.	grains
	373.242	grams
	0.37324	kilograms
	13.1657	ounces (avoir)
	12.	ounces (troy)
	240.	pennyweights
	0.822858	pounds (avoir)
	288.	scruples
	3.6735×10^{-4}	tons (long)
	3.7324×10^{-4}	tons (metric)
	4.1143×10^{-4}	tons (short)
pounds (mass)/acre-ft	3.6773×10^{-4}	kilograms/m^3
	0.02296	pounds/1000 ft^3
	2.2957×10^{-5}	pounds/ft^3
	1.3285×10^{-8}	pounds/in^3
	2.767×10^{-7}	tons (long)/yd^3
	1689.6	tons (short)/mi^3
pounds (mass)/acre-ft-day	0.02296	pounds/1000 ft^3-day
pounds (force)/ft^2	4.725×10^{-4}	atmospheres
	4.788×10^{-4}	bars
	0.035913	centimeters Hg at 0°C
	478.8	dynes/cm^2
	0.48824	grams/cm^2
	0.014139	inches Hg at 32°F
	4.8824×10^{-4}	kilograms/cm^2
	4.8824	kilograms/m^2
	4.8824×10^{-6}	kilograms/mm^2
	47.880	newtons/m^2
	0.111111	ounces/in^2
	47.880	pascals
	0.0069445	pounds/in^2
	4.464×10^{-4}	tons (long)/ft^2
	3.1002×10^{-6}	tons (long)/in^2
	0.0005	tons (short)/ft^2
	3.4722×10^{-6}	tons (short)/in^2

PROFESSIONAL PUBLICATIONS, INC. ● Belmont, CA

MULTIPLY	BY	TO OBTAIN
pounds (mass)/ft^3	0.26736	drams/ounce (U.S., liquid)
	7.31065	grains/ounce (U.S., liquid)
	7000.	grains/ft^3
	935.76	grains/gallon (U.S., liquid)
	0.016019	grams/cm^3
	16,019.	grams/m^3
	0.016019	grams/ml
	16.0185	kilograms/m^3
	0.00117821	mercury at 0°C
	2.1389	ounces/gallon (U.S., liquid)
	0.0092593	ounces (U.S., liquid)/in^3
	0.53472	ounces/quart (U.S., liquid)
	43,560.	pounds/acre-ft
	0.133681	pounds/gallon (U.S., liquid)
	5.7870×10^{-4}	pounds/in^3
	5.45415×10^{-9}	pounds/mil ft
	0.0310810	slugs/ft^3
	0.012054	tons (long)/yd^3
	7.3599×10^7	tons (short)/mi^3
pounds (mass)/1000 ft^3-day	43.56	pounds/acre-ft-day
	133.7	pounds/million gallons-day
pounds (mass)/ft-hr	0.4134	centipoises
	4.1338×10^{-3}	dyne-sec/cm^2
	0.004134	grams/cm-sec
	4.2135×10^{-5}	kilogram (force)-sec/m^2
	1.49	kilograms/m-hr
	0.004134	poises
	2.7778×10^{-4}	pounds (mass)/ft-sec
	8.634×10^{-6}	pound-sec/ft^2
	5.996×10^{-8}	pound-sec/in^2
	1.0	poundal-hr/ft^2
	2.778×10^{-4}	poundal-sec/ft^2
pounds (mass)/ft-sec	1488.	centipoises
	14.882	dyne-sec/cm^2
	14.88	grams/sec-cm
	5357.	kilograms/m-hr
	0.151	kilogram (force)-sec/m^2
	1.488	pascal-sec
	14.88	poises
	3600.	pounds (mass)/ft-hr
	0.03108	pound-sec/ft^2
	0.0002158	pound-sec/in^2
	3600.	poundal-hr/ft^2
	1.	poundal-sec/ft^2

MULTIPLY	BY	TO OBTAIN
pounds (mass)/gallon (U.S., liquid)	52,363.7	grains/ft^3
	0.119826	grams/cm^3
	119,826.	grams/m^3
	0.119826	grams/ml
	119.826	kilograms/m^3
	0.00881387	mercury at 0°C
	7.48052	pounds/ft^3
	0.00432900	pounds/in^3
	4.08000×10^{-8}	pounds/mil ft
	0.232502	slugs/ft^3
pounds H_2O/in	0.01335	miner's inches*
pounds (force)/in	175.126	newtons/m
pounds (force)/in^2	0.068046	atmospheres
	0.06895	bars
	5.1715	centimeters Hg at 0°C
	68,948.	dynes/cm^2
	2.307	feet H_2O
	70.307	grams/cm^2
	0.703	grams/mm^2
	27.7	inches H_2O
	2.036	inches Hg at 32°F
	0.070307	kilograms/cm^2
	703.07	kilograms/m^2
	7.0307×10^{-4}	kilograms/mm^2
	6.895	kilopascals
	0.006895	megapascals
	6895.	newtons/m^2
	16.	ounces/in^2
	6894.8.	pascals
	144.	pounds/ft^2
	0.064286	tons (long)/ft^2
	4.464×10^{-4}	tons (long)/in^2
	0.072	tons (short)/ft^2
	0.0005	tons (short)/in^2

* ID, KS, ND, NE, NM, NV, SD, UT

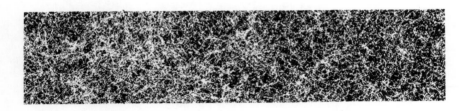

PROFESSIONAL PUBLICATIONS, INC. ● Belmont, CA

MULTIPLY	BY	TO OBTAIN
pounds (mass)/in^3	462.	drams/ounce (U.S., liquid)
	1.20960×10^7	grains/ft^3
	12,633.	grains/ounce (U.S., liquid)
	27.680	grams/cm^3
	2.7680×10^7	grams/m^3
	27.680	grams/ml
	27,680.	kilograms/m^3
	2.036	mercury at 0°C
	3696.0	ounces/gallon (U.S., liquid)
	16.	ounces (U.S., liquid)/in^3
	924.	ounces/quart (U.S., liquid)
	7.5272×10^7	pounds/acre-ft
	1728.	pounds/ft^3
	231.	pounds/gallon (U.S., liquid)
	9.42478×10^{-6}	pounds/mil foot
	53.7079	slugs/ft^3
	1.2718×10^{11}	tons (short)/mi^3
pounds (mass)/mil ft	1.28343×10^{12}	grains/ft^3
	2.9369×10^6	grams/cm^3
	2.9369×10^{12}	grams/m^3
	2.9369×10^6	grams/ml
	2.9369×10^9	kilograms/m^3
	216,025.	mercury at 0°C
	1.83347×10^8	pounds/ft^3
	2.45099×10^7	pounds/gallon
	106,103.	pounds/in^3
	5.69858×10^6	slugs/ft^3
pounds (mass)/million gallons	0.1198	milligrams/ℓ
pounds (mass)/million gallons-day	0.00748	pounds/1000 ft^3-day
pounds (mass) H$_2$O/min	7.5594	cubic centimeters/sec
	0.01602	cubic feet/min
	2.670×10^{-4}	cubic feet/sec
	5.933×10^{-4}	cubic yards/min
	0.11982	gallons (U.S., liquid)/min
	0.0019971	gallons (U.S., liquid)/sec
	0.45366	liters/min
	0.007559	liters/sec

MULTIPLY	BY	TO OBTAIN
pounds (mass)/ton (long)	453.596	grams/ton (long)
	446.429	grams/ton (metric)
	436.392	grams/ton (short)
	0.095663	karats
	0.44643	milligrams/g
	446.43	milligrams/kg
	13.024	milligrams/ton (assay)
	16.	ounces/ton (long)
	14.286	ounces/ton (short)
	0.89286	pounds/ton (short)
pounds (mass)/ton (short)	508.027	grams/ton (long)
	500.	grams/ton (metric)
	488.76	grams/ton (short)
	0.10714	karats
	0.5	milligrams/g
	500.	milligrams/kg
	14.588	milligrams/ton (assay)
	17.920	ounces/ton (long)
	16.	ounces/ton (short)
	1.12	pounds/ton (long)
pound (force)-ft	1.3558×10^7	dyne-cm
	0.13826	kilogram-m
	12.	pound-in
	32.174	poundal-ft
pound (mass)-ft^2	421,401.	gram-cm^2
	421.4	kilogram-cm^2
	144.	pound-in^2
pound (force)-in	1.1299×10^6	dyne-cm
	0.011521	kilogram-m
	0.083333	pound-ft
	2.6812	poundal-ft
pound (mass)-in^2	2926.4	gram-cm^2
	2.9264	kilogram-cm^2
	0.006945	pound-ft^2

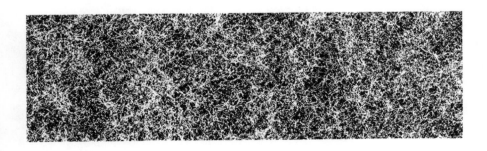

PROFESSIONAL PUBLICATIONS, INC. ● Belmont, CA

MULTIPLY	BY	TO OBTAIN
pound (force)-sec/ft^2	47,880.	centipoises
	478.8	dyne-sec/cm^2
	0.48824	gram (force)-sec/cm^2
	478.8	grams/cm-sec
	4.882	kilogram (force)-sec/m^2
	172,369.	kilograms/m-hr
	47.88	pascal-sec
	478.8	poises
	115,827.	pounds (mass)/ft-hr
	32.174	pounds (mass)/ft-sec
	0.0069444	pound-sec/in^2
	115,826.	poundal-hr/ft^2
	32.174	poundal-sec/ft^2
pound (force)-sec/in^2	6.895×10^6	centipoises
	68,950.	dyne-sec/in^2
	70.307	gram (force)-sec/cm^2
	68,948.	grams/cm-sec
	703.	kilogram (force)-sec/m^2
	2.48×10^7	kilograms/m-hr
	68,948.	poises
	1.668×10^7	pounds (mass)/ft-hr
	4633.	pounds (mass)/ft-sec
	144.	pound-sec/ft^2
	1.668×10^7	poundal-hr/ft^2
	4633.	poundal-sec/ft^2
poundals	1.3825	crinals
	13,826.	dynes
	217.57	grains
	14.100	grams (force)
	0.01410	kilograms (force)
	3.1081×10^{-5}	kips
	0.13826	newtons
	0.031081	pounds
	1.383×10^{-4}	sthènes
	1.3875×10^{-5}	tons (long)
	1.5541×10^{-5}	tons (short)
poundals/ft^2	1.4292×10^{-5}	atmospheres
	1.448×10^{-5}	bars
	1.4482	pascals
	0.030246	pounds/ft^2
	2.1004×10^{-4}	pounds/in^2

PROFESSIONAL PUBLICATIONS, INC. • Belmont, CA

MULTIPLY	BY	TO OBTAIN
poundal-ft	421,401.	dyne-cm
	0.0042971	kilogram-m
	0.031081	pound-ft
	0.37297	pound-in
poundal-hr/ft^2	0.41334	centipoises
	0.0041334	dyne-sec/cm^2
	0.0041334	grams/cm-sec
	4.214×10^{-5}	kilogram (force)-sec/m^2
	1.49	kilograms/m-hr
	0.0041334	poises
	2.7778×10^{-4}	pounds (mass)/ft-sec
	1.	pounds (mass)/ft-hr
	8.634×10^{-6}	pound-sec/ft^2
	5.995×10^{-8}	pound-sec/in^2
	3600.	poundal-sec/ft^2
poundal-sec/ft^2	1488.	centipoises
	0.001488	dyne-sec/cm^2
	14.88	grams/cm-sec
	0.152	kilogram (force)-sec/m^2
	5357.	kilograms/m-hr
	14.88	poises
	1.	pounds (mass)/ft-sec
	3600.	pounds (mass)/ft-hr
	0.031081	pound-sec/ft^2
	2.158×10^{-4}	pound-sec/in^2
	2.7778×10^{-4}	poundal-hr/ft^2
ppm	(see milligrams/ℓ)	

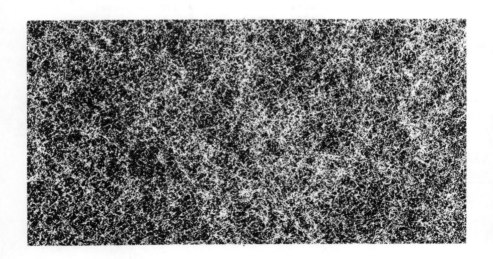

MULTIPLY	BY	TO OBTAIN
quadrants	90.	degrees
	100.	grades
	5400.	minutes
	1.5708	radians
	0.25	revolutions
	324,000.	seconds
	3.	signs
quarts (U.S., dry)	0.03125	bushels
	1101.2	cubic centimeters
	0.03889	cubic feet
	67.2	cubic inches
	0.0011012	cubic meters
	0.0014403	cubic yards
	0.25	gallons (U.S., dry)
	0.29091	gallons (U.S., liquid)
	0.125	pecks
	2.	pints (U.S., dry)
	0.0011012	steres
quarts (U.S., liquid)	0.0079365	barrels (31.5 U.S. gallons)
	0.0059524	barrels (42 U.S. gallons)
	946.35	cubic centimeters
	0.033420	cubic feet
	57.750	cubic inches
	9.4635×10^{-4}	cubic meters
	0.0012378	cubic yards
	256.	drams (U.S., liquid)
	0.027778	firkins
	0.21484	gallons (U.S., dry)
	0.25	gallons (U.S., liquid)
	8.	gills
	0.0039683	hogsheads
	0.946353	liters
	946.35	milliliters
	15,360.	minims
	32.	ounces (U.S., liquid)
	2.	pints (U.S., liquid)
quarters (cloth)	22.86	centimeters
	0.2	ells
	0.75	feet
	9.	inches

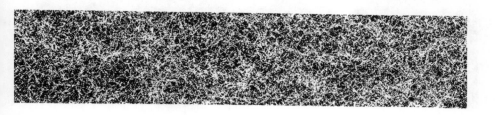

MULTIPLY	BY	TO OBTAIN
quarters (long)	5.6	centals
	5.	hundredweights (long)
	5.6	hundredweights (short)
	254.01	kilograms
	8960.	ounces (avoir)
	560.	pounds (avoir)
	1.12	quarters (short)
	2.5401	quintals
	17.405	slugs
	0.25	tons (long)
	0.25401	tons (metric)
	0.28	tons (short)
quarters (short)	5.	centals
	4.4643	hundredweights (long)
	5.	hundredweights (short)
	226.796	kilograms
	8000.	ounces (avoir)
	500.	pounds (avoir)
	0.89286	quarters (long)
	2.26796	quintals
	15.541	slugs
	0.22321	tons (long)
	0.22680	tons (metric)
	0.25	tons (short)
quintals	2.2046	centals
	1.9684	hundredweights (long)
	2.2046	hundredweights (short)
	100.	kilograms
	3527.4	ounces (avoir)
	3215.1	ounces (troy)
	220.46	pounds (avoir)
	267.92	pounds (troy)
	0.39369	quarters (long)
	0.44093	quarters (short)
	6.85219	slugs
	0.098421	tons (long)
	0.1	tons (metric)
	0.11023	tons (short)

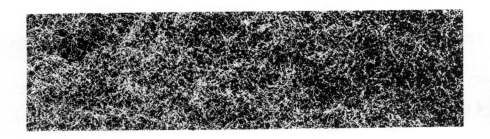

PROFESSIONAL PUBLICATIONS, INC. • Belmont, CA

MULTIPLY	BY	TO OBTAIN
radians	57.2958	degrees
	57° 17′ 44.8″	deg/min-sec
	63.662	grades
	3437.75	minutes
	0.63662	quadrants
	0.159155	revolutions
	206,265.	seconds
	1.90986	signs
radians/cm	57.296	degrees/cm
	1746.4	degrees/ft
	145.5	degrees/in
	3437.7	minutes/cm
radians/sec	57.296	degrees/sec
	13,751.	revolutions/day
	9.549	revolutions/min
	0.1592	revolutions/sec
radians/sec^2	572.96	revolutions/min^2
	9.5493	revolutions/min-sec
	0.159155	revolutions/sec^2
rads	0.01	grays
rambles	1320.	feet
refrigeration tons	11,991.	BTU/hr
	100.	cubic feet
	3516.8	watts
register tons	2.8317	cubic meters
rems	0.01	sieverts
revolutions	360.	degrees
	400.	grades
	21,600.	minutes
	4.	quadrants
	6.28319	radians
	1.296×10^6	seconds
	12.	signs
	0.33333	trikes
revolutions/day	0.00416667	degrees/sec
	7.2722×10^{-5}	radians/sec
	6.9444×10^{-4}	revolutions/min
	1.1574×10^{-5}	revolutions/sec
revolutions/min	6.	degrees/sec
	0.10472	radians/sec
	1440.	revolutions/day
	0.016667	revolutions/sec

PROFESSIONAL PUBLICATIONS, INC. ● Belmont, CA

MULTIPLY	BY	TO OBTAIN
revolutions/min^2	0.0017453	radians/sec^2
	0.0166667	revolutions/min-sec
	2.77778×10^{-4}	revolutions/sec^2
revolutions/min-sec	0.104720	radians/sec^2
	60.	revolutions/min^2
	0.0166667	revolutions/sec^2
revolutions/sec	360.	degrees/sec
	6.2832	radians/sec
	86,400.	revolutions/day
	60.	revolutions/min
revolutions/sec^2	6.28319	radians/sec^2
	3600.	revolutions/min^2
	60.	revolutions/min-sec
reyns	6.8948×10^6	centipoises
rods	502.921	centimeters
	0.25	chains (Gunter's or surveyor's)
	0.165	chains (Ramsden's or engineer's)
	2.75	fathoms
	16.5	feet
	0.025	furlongs
	198.	inches
	0.0050292	kilometers
	0.00104167	leagues (statute)
	25.	links (Gunter's or surveyor's)
	16.5	links (Ramsden's or engineer's)
	5.0292	meters
	6.6	paces
	1.0	perches
	1.0	poles
	5.5	yards
roentgens	2.58×10^{-4}	coulombs/kg
	83.8	ergs absorbed/g air
roods	0.25	acres
	10.117	ares
	0.10117	hectares
	2.5	square chains (Gunter's or surveyor's)
	10,890.	square feet
	1012.	square meters
	40.	square rods
	1210.	square yards
ropes	20.	feet
	240.	inches
	6.096	meters
	6.6667	yards
rpm	(see revolutions/minute)	

MULTIPLY	BY	TO OBTAIN
scruples	6.480	carats
	0.73143	drams (avoir)
	0.33333	drams (troy)
	20.	grains
	1.2960	grams
	0.0012960	kilograms
	0.045714	ounces (avoir)
	0.0416667	ounces (troy)
	0.83333	pennyweights
	0.0028571	pounds (avoir)
	0.0034722	pounds (troy)
	0.044434	tons (assay)
seconds (angular)	2.77778×10^{-4}	degrees
	3.0864×10^{-4}	grades
	0.0166667	minutes
	3.08642×10^{-6}	quadrants
	4.84814×10^{-6}	radians
	7.71605×10^{-7}	revolutions
	9.2593×10^{-6}	signs
seconds (sidereal)	1.1574×10^{-5}	days (sidereal)
	1.1542×10^{-5}	days (solar)
	2.7702×10^{-4}	hours (solar)
	0.016621	minutes (solar)
	3.9087×10^{-7}	months (lunar)
	3.7948×10^{-7}	months (mean)
	0.99727	seconds (solar)
	1.6489×10^{-6}	weeks
	3.1537×10^{-8}	years (leap)
	3.1601×10^{-8}	years (sidereal)
	3.1602×10^{-8}	years (solar)
seconds (solar)	1.1606×10^{-5}	days (sidereal)
	1.1574×10^{-5}	days (solar)
	2.7778×10^{-4}	hours (solar)
	0.016667	minutes (solar)
	3.9194×10^{-7}	months (lunar)
	3.8052×10^{-7}	months (mean)
	1.0027	seconds (sidereal)
	1.0×10^{8}	shakes
	1.6534×10^{-6}	weeks
	3.1623×10^{-8}	years (leap)
	3.1688×10^{-8}	years (sidereal)
	3.1689×10^{-8}	years (solar)

PROFESSIONAL PUBLICATIONS, INC. • Belmont, CA

MULTIPLY	BY	TO OBTAIN
sections	640.	acres
	259.	hectares
	2.590	square kilometers
	1.	square miles
	102,400.	square rods
	0.027778	townships
shakes	0.01	microseconds
	1.0×10^{-8}	seconds
sheds	1.0×10^{-52}	square meters
shipping tons	40.	cubic feet
siemens	1.0×10^{-9}	abmhos
	1.	mhos
	8.988×10^{11}	statmhos
sieverts	100.	vems
signs	30.	degrees
	33.333	grades
	1800.	minutes
	0.33333	quadrants
	0.52360	radians
	0.083333	revolutions
	108,000.	seconds
skeins	3.	bolts
	96.	ells
	360.	feet
	4320.	inches
	0.109728	kilometers
	109.728	meters
	120.	yards
slugs	0.32174	centals
	1.	geepounds
	14,594.	grams
	0.28727	hundredweights (long)
	0.32174	hundredweights (short)
	14.594	kilograms
	514.78	ounces (avoir)
	1.	pounds (force)-sec^2/ft
	32.174	pounds (mass)
	0.057454	quarters (long)
	0.064348	quarters (short)
	0.14594	quintals
	0.014363	tons (long)
	0.014594	tons (metric)
	0.016087	tons (short)

MULTIPLY	BY	TO OBTAIN
slugs/ft^3	225,219.	grains/ft^3
	515,379.	grains/m^3
	0.51538	grams/cm^3
	0.51538	grams/ml
	515.379	kilograms/m^3
	0.0379105	mercury at 0°C
	32.174	pounds/ft^3
	4.3011	pounds/gallon
	0.018619	pounds/in^3
	1.7548×10^{-7}	pounds/mil ft
slugs/ft-sec	47.88	pascal-sec
spans	22.86	centimeters
	12.	digits
	0.75	feet
	2.25	hands
	9.	inches
	3.	palms
spheres	2.	hemispheres
	720.	spherical degrees
	8.	spherical right angles
	41,253.	square degrees
	12.5664	steradians
	1.	steregons
spherical degrees	0.0027778	hemispheres
	0.00138889	spheres
	0.0055556	spherical right angles
	57.296	square degrees
	0.017454	steradians
	0.00138889	steregons
spherical right angles	0.25	hemispheres
	0.125	spheres
	90.	spherical degrees
	5156.62	square degrees
	1.5708	steradians
	0.125	steregons

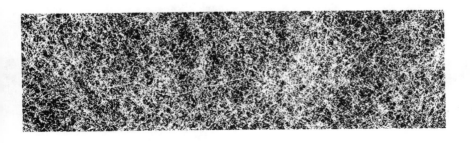

MULTIPLY	BY	TO OBTAIN
square centimeters	2.47104×10^{-8}	acres
	1.0×10^{-6}	ares
	1.2732	circular centimeters
	0.0013705	circular feet
	0.197353	circular inches
	12.732	circular millimeters
	197,353.	circular mils
	2.47105×10^{-7}	square chains (Gunter's or surveyor's)
	0.0010764	square feet
	0.155	square inches
	0.00247105	square links (Gunter's or surveyor's)
	1.0×10^{-4}	square meters
	100.	square millimeters
	155,000.	square mils
	1.196×10^{-4}	square yards
square centimeters/day	0.001	liters/cm-day
	1.1574×10^{-5}	square centimeters/sec
	1.794×10^{-6}	square inches/sec
square centimeters/sec	100.	centistokes
	86.4	liters/cm-day
	1.	poise-cm^3/g
	0.016017	poise-ft^3/lbm
	0.061024	poise-in^3/g
	86,400.	square centimeters/day
	3.875	square feet/hr
	0.0010764	square feet/sec
	0.1550	square inches/sec
	0.36	square meters/hr
	1.0×10^{-4}	square meters/sec
	1.	stokes
square chains (Gunter's or surveyor's)	0.1	acres
	4.04686	ares
	404.686	centares
	0.0404686	hectares
	0.4	roods
	4.04686×10^6	square centimeters
	4356.	square feet
	627,264.	square inches
	4.04686×10^{-4}	square kilometers
	10,000.	square links (Gunter's or surveyor's)
	404.686	square meters
	1.5625×10^{-4}	square miles
	16.	square perches
	16.	square rods
	484.	square yards
	4.34026×10^{-6}	townships

PROFESSIONAL PUBLICATIONS, INC. ● Belmont, CA

MULTIPLY	BY	TO OBTAIN
square degrees	4.84814×10^{-5}	hemispheres
	2.42407×10^{-5}	spheres
	0.017454	spherical degrees
	1.93926×10^{-4}	spherical right angles
	3.0462×10^{-4}	steradians
	2.424×10^{-5}	steregons
square feet	2.29568×10^{-5}	acres
	9.29034×10^{-4}	ares
	0.0929034	centares
	1.2732	circular feet
	183.347	circular inches
	11,828.84	circular millimeters
	1.83347×10^{8}	circular mils
	10^{-26}	doors
	9.29034×10^{-6}	hectares
	9.183×10^{-5}	roods
	929.034	square centimeters
	2.29569×10^{-4}	square chains (Gunter's or surveyor's)
	144.	square inches
	9.29034×10^{-8}	square kilometers
	2.29569	square links (Gunter's or surveyor's)
	0.0929034	square meters
	3.58702×10^{-8}	square miles
	92,903.	square millimeters
	1.44×10^{10}	square mils
	0.0036731	square perches
	0.0036731	square rods
	0.1296	square varas
	0.111112	square yards
	9.96391×10^{-10}	townships
square feet/hr	25.806	centistokes
	0.25806	square centimeters/sec
	2.778×10^{-4}	square feet/sec
	0.04	square inches/sec
	0.092903	square meters/hr
	2.5806×10^{-5}	square meters/sec
	0.25806	stokes
square feet/sec	92,900.	centistokes
	929.	square centimeters/sec
	3600.	square feet/hr
	144.	square inches/sec
	334.45	square meters/hr
	0.09290	square meters/sec
	929.	stokes

PROFESSIONAL PUBLICATIONS, INC. ● Belmont, CA

MULTIPLY	BY	TO OBTAIN
square inches	1.59422×10^{-7}	acres
	6.4516×10^{-6}	ares
	6.4516×10^{24}	barns
	6.4516×10^{-4}	centares
	8.2145	circular centimeters
	0.008842	circular feet
	1.27324	circular inches
	82.1443	circular millimeters
	1.27324×10^{6}	circular mils
	6.4516×10^{-8}	hectares
	6.4516	square centimeters
	1.59423×10^{-6}	square chains (Gunter's or surveyor's)
	0.0069444	square feet
	6.4516×10^{-10}	square kilometers
	0.0159423	square links (Gunter's or surveyor's)
	6.4516×10^{-4}	square meters
	2.49098×10^{-10}	square miles
	645.16	square millimeters
	1.0×10^{6}	square mils
	2.55076×10^{-5}	square perches
	2.55076×10^{-5}	square rods
	7.71605×10^{-4}	square yards
square inches/sec	645.2	centistokes
	557.42	liters/cm-day
	6.4516	poise-cm^3/gravity
	0.10333	poise-ft^3/lbm
	0.39366	poise-in^3/gravity
	557,420.	square centimeters/day
	6.4516	square centimeters/sec
	25.	square feet/hr
	0.0069444	square feet/sec
	2.3226	square meters/hr
	6.452×10^{-4}	square meters/sec
	6.452	stokes
square kilometers	247.104	acres
	10,000.	ares
	1.0×10^{6}	centares
	100.	hectares
	0.3861	sections
	2471.1	square chains (Gunter's or surveyor's)
	1.07640×10^{7}	square feet
	2.47105×10^{7}	square links (Gunter's or surveyor's)
	1.0×10^{6}	square meters
	0.386102	square miles
	39,536.9	square perches
	39,536.7	square rods
	1.19599×10^{6}	square yards
	0.0107250	townships

MULTIPLY	BY	TO OBTAIN
square links	1.0×10^{-6}	acres
(Gunter's or	4.0469×10^{-4}	ares
surveyor's)	0.040469	centares
	404.69	square centimeters
	1.0×10^{-4}	square chains (Gunter's or surveyor's)
	0.4356	square feet
	62.726	square inches
	0.040469	square meters
	0.0016	square rods (also square perches)
	0.0484	square yards
square meters	2.47104×10^{-4}	acres
	0.01	ares
	1.0×10^{28}	barns
	1.	centares
	12,732.	circular centimeters
	13.705	circular feet
	1973.53	circular inches
	1.27324×10^{5}	circular millimeters
	1.97353×10^{9}	circular mils
	1.0×10^{-4}	hectares
	9.8842×10^{-4}	roods
	10,000.	square centimeters
	0.00247105	square chains (Gunter's or surveyor's)
	10.7639	square feet
	1550.	square inches
	1.0×10^{-6}	square kilometers
	24.7105	square links (Gunter's or surveyor's)
	3.861×10^{-7}	square miles
	1.0×10^{6}	square millimeters
	0.039537	square perches
	0.039537	square rods
	1.395	square varas
	1.19599	square yards
	1.07250×10^{-8}	townships
square meters/hr	277.78	centistokes
	2.7778	square centimeters/sec
	10.764	square feet/hr
	0.00299	square feet/sec
	0.43056	square inches/sec
	2.778×10^{-4}	square meters/sec
	2.7778	stokes
square meters/sec	1.0×10^{6}	centistokes
	10,000.	square centimeters/sec
	38,750.	square feet/hr
	10.764	square feet/sec
	1550.	square inches/sec
	3600.	square meters/hr
	1.0×10^{4}	stokes

MULTIPLY	BY	TO OBTAIN
square miles	640.	acres
	25,900.	ares
	2.59000×10^6	centares
	259.	hectares
	1.	sections
	6400.	square chains (Gunter's or surveyor's)
	2.78785×10^7	square feet
	4.01451×10^9	square inches
	2.59000	square kilometers
	6.4×10^7	square links (Gunter's or surveyor's)
	2.59000×10^6	square meters
	102,400.	square perches
	102,400.	square rods
	3.09761×10^6	square yards
	0.0277778	townships
square mile-inch	53.3	acre-ft
square mile-inch/day	26.88	cubic feet/sec
square millimeters	1.0×10^{22}	barns
	0.00197353	circular inches
	1.2732	circular millimeters
	1973.5	circular mils
	0.01	square centimeters
	1.07639×10^{-5}	square feet
	0.00155	square inches
	1.0×10^{-6}	square meters
	1550.	square mils
	1.19599×10^{-6}	square yards
square mils	1.2732×10^{-6}	circular inches
	8.21418×10^{-4}	circular millimeters
	1.2732	circular mils
	6.45161×10^{-6}	square centimeters
	6.94444×10^{-9}	square feet
	1.0×10^{-6}	square inches
	6.45161×10^{-10}	square meters
	6.45161×10^{-4}	square millimeters

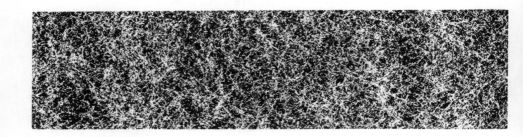

MULTIPLY	BY	TO OBTAIN
square perches	0.00625	acres
	0.252929	ares
	25.2929	centares
	0.00252929	hectares
	0.0625	square chains (Gunter's or surveyor's)
	272.25	square feet
	39,204.	square inches
	2.52929×10^{-5}	square kilometers
	625.	square links (Gunter's or surveyor's)
	25.293	square meters
	9.76563×10^{-6}	square miles
	1.	square rods
	30.25	square yards
	2.71266×10^{-7}	townships
square rods	0.00625	acres
	0.252929	ares
	25.2930	centares
	0.00252930	hectares
	0.025	roods
	9.766×10^{-6}	sections
	0.0625	square chains (Gunter's or surveyor's)
	272.251	square feet
	39,204.2	square inches
	2.52930×10^{-5}	square kilometers
	625.	square links (Gunter's or surveyor's)
	25.2930	square meters
	9.76566×10^{-6}	square miles
	1.	square perches
	30.25	square yards
	2.71267×10^{-7}	townships
square varas	1.	ferrados
	7.716	square feet
	0.7168	square meters
	0.8573	square yards

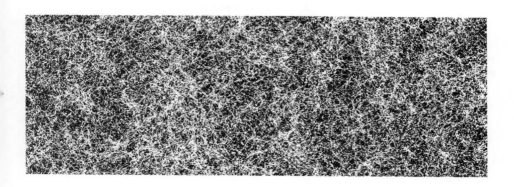

PROFESSIONAL PUBLICATIONS, INC. ● Belmont, CA

MULTIPLY	BY	TO OBTAIN
square yards	2.06611×10^{-4}	acres
	0.00836128	ares
	0.836127	centares
	1650.12	circular inches
	1.06459×10^5	circular millimeters
	1.65012×10^9	circular mils
	8.36127×10^{-5}	hectares
	8.265×10^{-4}	roods
	8361.27	square centimeters
	0.00206612	square chains (Gunter's or surveyor's)
	9.	square feet
	1296.	square inches
	8.36127×10^{-7}	square kilometers
	20.6612	square links (Gunter's or surveyor's)
	0.836127	square meters
	3.22831×10^{-7}	square miles
	836,127.	square millimeters
	1.29600×10^{12}	square mils
	0.0330579	square perches
	0.0330577	square rods
	1.166	square varas
	8.96748×10^{-9}	townships
standard cubic feet	(see cubic feet)	
statamperes	3.3356×10^{-11}	abamperes
	3.3356×10^{-10}	amperes
	3.45667×10^{-15}	faradays/sec
	3.3356×10^{-7}	milliamperes
statamperes/cm^2	3.335×10^{-11}	abamperes/cm^2
	3.335×10^{-10}	amperes/cm^2
	2.151×10^{-9}	amperes/in^2
	3.335×10^{-6}	amperes/m^2
statcoulombs	3.3356×10^{-11}	abcoulombs
	9.2657×10^{-14}	ampere-hr
	3.3356×10^{-10}	coulombs
	2.0947×10^9	electronic charges
	2.7906×10^{-4}	electrostatic ft-lbf-sec
	3.4558×10^{-15}	faradays
statcoulombs/cm^2	3.335×10^{-11}	abcoulombs/cm^2
	3.335×10^{-9}	coulombs/cm^2
	2.151×10^{-9}	coulombs/in^2
	3.335×10^{-5}	coulombs/m^2

MULTIPLY	BY	TO OBTAIN
statfarads	1.11265×10^{-21}	abfarads
	1.11265×10^{-12}	farads
	1.11265×10^{-6}	microfarads
	1.11265	micromicrofarads
	1.11265	picofarads
stathenrys	8.98755×10^{20}	abhenrys
	8.98755×10^{11}	henrys
	8.98755×10^{17}	microhenrys
	8.98755×10^{14}	millihenrys
statmhos	1.1127×10^{-12}	mhos
	1.1127×10^{-12}	siemens
statoersteds	3.336×10^{-11}	oersteds
statohms	8.9876×10^{20}	abohms
	$898{,}755.$	megohms
	8.9876×10^{17}	microhms
	8.9876×10^{11}	ohms
statvolts	2.99793×10^{10}	abvolts
	299.694	international volts
	299.793	volts
statwebers	2.998×10^{10}	gauss-cm^2
	2.998×10^{10}	maxwells
	299.8	webers
steradians	0.15916	hemispheres
	0.07958	spheres
	57.2958	spherical degrees
	0.6366	spherical right angles
	3282.8	square degrees
	0.079578	steregons
steregons	$2.$	hemispheres
	$1.$	spheres
	$720.$	spherical degrees
	$8.$	spherical right angles
	$41{,}253.$	square degrees
	12.5664	steradians

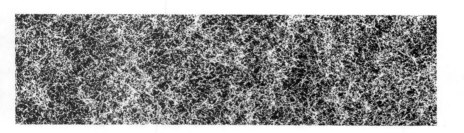

PROFESSIONAL PUBLICATIONS, INC. • Belmont, CA

MULTIPLY	BY	TO OBTAIN
steres	423.78	board-ft
	28.378	bushels
	0.78827	chaldrons
	2.2072	cord-ft
	0.27590	cords
	1.0×10^6	cubic centimeters
	35.315	cubic feet
	61,024.	cubic inches
	1.	cubic meters
	1.3080	cubic yards
	227.	gallons (U.S., dry)
	264.2	gallons (U.S., liquid)
	113.51	pecks
	1816.2	pints (U.S., dry)
	908.08	quarts (U.S., dry)
sthènes	10,000.	crinals
	1.0×10^8	dynes
	101.972	kilograms (force)
	0.22481	kips
	1000.	newtons
	224.8	pounds
	7233.	poundals
	0.1004	tons (long)
	0.1124	tons (short)
stigmas	0.01	angstroms
	3.28084×10^{-12}	feet
	3.9370×10^{-11}	inches
	1.0×10^{-12}	meters
	1.	micromicrons
	1.0×10^{-6}	microns
	1.	millionth microns
stilbs	31,415.9	apostilbs
	1.	candles/cm^2
	929.031	candles/ft^2
	6.45157	candles/in^2
	10,000.	candles/m^2
	2918.6	foot-lamberts
	3.14159	lamberts
	1.	lumens/cm^2-sterad
	929.03	lumens/ft^2-sterad
	3141.59	millilamberts

PROFESSIONAL PUBLICATIONS, INC. • Belmont, CA

MULTIPLY	BY	TO OBTAIN
stokes	100.	centistokes
	1.	square centimeters/sec
	3.875	square feet/hr
	0.0010764	square feet/sec
	0.1550	square inches/sec
	0.3600	square meters/hr
	1.0×10^{-4}	square meters/sec
stones	6.350	kilograms
	14.	pounds

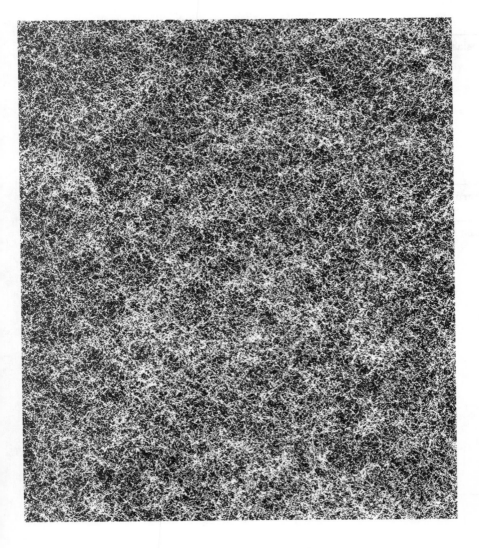

MULTIPLY	BY	TO OBTAIN
tablespoons	14.787	cubic centimeters
	4.	drams (U.S., liquid)
	0.5	ounces (U.S., liquid)
	3.	teaspoons
teaspoons	4.929	cubic centimeters
	1.333	drams (U.S., liquid)
	0.1667	ounces (U.S., liquid)
	0.33333	tablespoons
tenthmeters	1.	angstroms
	3.28084×10^{-10}	feet
	3.9370×10^{-9}	inches
	1.0×10^{-10}	meters
teslas	7958.	ampere-turns/cm
	20,213.	ampere-turns/in
	1.0×10^{9}	gammas
	10,000.	gauss
	10,000.	gilberts/cm
	10,000.	lines/cm^2
	64,520.	lines/in^2
	1.0×10^{-4}	webers/cm^2
	6.452×10^{-4}	webers/in^2
	1.	webers/m^2
therms	1.0×10^{5}	BTU
	1.05506×10^{8}	joules
tons (long)	22.4	centals
	9.964×10^{8}	dynes
	1.5680×10^{7}	grains
	1.0161×10^{6}	grams (mass)
	20.	hundredweights (long)
	22.4	hundredweights (short)
	1016.1	kilograms (mass)
	9964.	newtons
	35,840.	ounces (avoir)
	32,666.7	ounces (troy)
	2240.	pounds (avoir)
	2722.22	pounds (troy)
	72,070.	poundals
	4.	quarters (long)
	4.48	quarters (short)
	10.161	quintals
	69.62	slugs (mass)
	9.964	sthènes
	1.01605	tons (metric)
	1.12	tons (short)

MULTIPLY	BY	TO OBTAIN
tons (metric)	22.046	centals
	1.543×10^7	grains
	1.0×10^6	grams
	19.684	hundredweights (long)
	22.046	hundredweights (short)
	1000.	kilograms
	35,274.	ounces (avoir)
	32,150.7	ounces (troy)
	2204.62	pounds (avoir)
	2679.23	pounds (troy)
	3.93685	quarters (long)
	4.40927	quarters (short)
	10.	quintals
	68.522	slugs
	1.	tonnes
	0.98421	tons (long)
	1.1023	tons (short)
tons (refrigeration)	12,000.	BTU/hr
	200.	BTU/min
	3517.	watts
tons (short)	20.	centals
	8.8964×10^8	dynes
	1.40×10^7	grains
	907,185.	grams (mass)
	17.857	hundredweights (long)
	20.	hundredweights (short)
	907.185	kilograms
	8896.4	newtons
	32,000.	ounces (avoir)
	29,166.66	ounces (troy)
	2000.	pounds (avoir)
	2430.55	pounds (troy)
	64,348.	poundals
	3.57145	quarters (long)
	4.	quarters (short)
	9.07185	quintals
	62.162	slugs (mass)
	8.896	sthènes
	0.89286	tons (long)
	0.907185	tons (metric)
tons (long)/ft^2	1.0585	atmospheres
	1.0725	bars
	1.0937	kilograms/cm^2
	248.89	ounces/in^2
	2240.	pounds/ft^2
	15.5556	pounds/in^2
	0.0069444	tons (long)/in^2
	1.12	tons (short)/ft^2
	0.0077780	tons (short)/in^2

PROFESSIONAL PUBLICATIONS, INC. • Belmont, CA

MULTIPLY	BY	TO OBTAIN
tons (short)/ft^2	0.94508	atmospheres
	0.95761	bars
	71.826	centimeters Hg at 0°C
	957,605.	dynes/cm^2
	976.49	grams (mass)/cm^2
	0.97649	kilograms/cm^2
	9764.9	kilograms/m^2
	0.0097649	kilograms/mm^2
	222.22	ounces/in^2
	2000.	pounds/ft^2
	13.889	pounds/in^2
	0.89286	tons (long)/ft^2
	0.0062004	tons (long)/in^2
	0.0069444	tons (short)/in^2
tons (short)/ft^3	2000.	pounds/ft^3
	267.	pounds/gallon
	1.157	pounds/in^3
	27.	tons (short)/yd^3
tons (long)/in^2	152.42	atmospheres
	154.44	bars
	11,584.	centimeters Hg at 0°C
	4561.	inches Hg at 32°F
	1.5749	kilograms/mm^2
	322,561.	pounds/ft^2
	2240.	pounds/in^2
	144.	tons (long)/ft^2
	161.28	tons (short)/ft^2
	1.12	tons (short)/in^2
tons (short)/in^2	136.092	atmospheres
	137.90	bars
	1.3790×10^8	dynes/cm^2
	140,614.	grams (force)/cm^2
	140.61	kilograms/cm^2
	1.4061×10^6	kilograms/m^2
	1.4061	kilograms/mm^2
	32,000.	ounces/in^2
	288,000.	pounds/ft^2
	2000.	pounds/in^2
	128.57	tons (long)/ft^2
	0.89286	tons (long)/in^2
	144.	tons (short)/ft^2

MULTIPLY	BY	TO OBTAIN
tons (metric)/m^3	1000.	grams/ℓ
	1000.	kilograms/m^3
	2.719	pounds/acre-ft
	62.428	pounds/ft^3
	0.7525	tons (long)/yd^3
	4595.	tons (short)/mi^3
tons (short)/mi^3	2.1765×10^{-7}	kilograms/m^3
	5.9186×10^{-4}	pounds/acre-ft
	1.3587×10^{-8}	pounds/ft^3
	7.863×10^{-12}	pounds/in^3
	1.638×10^{-10}	tons (long)/yd^3
tons (long)/yd^3	1.3289	grams/cm^3
	1328.9	kilograms/m^3
	3.614×10^6	pounds/acre-ft
	82.963	pounds/ft^3
	11.09	pounds/gallon
	0.0401	pounds/in^3
	2240.	pounds/yd^3
	1.3289	tons (metric)/m^3
	6.106×10^9	tons (short)/mi^3
tons (short)/yd^3	1.187	grams/cm^3
	1187.	kilograms/m^3
	74.07	pounds/ft^3
	9.902	pounds/gallon
	0.04287	pounds/in^3
	0.8929	tons (long)/yd^3
	1.187	tons (metric)/m^3
	0.037	tons (short)/ft^3
torrs	0.0013158	atmospheres
	0.0013332	bars
	1.	millimeters Hg at 0°C
	133.322	pascals

PROFESSIONAL PUBLICATIONS, INC. ● Belmont, CA

MULTIPLY	BY	TO OBTAIN
townships	23,040.	acres
	932,399.	ares
	9323.99	hectares
	36.	sections
	230,401.	square chains (Gunter's or surveyor's)
	93.240	square kilometers
	9.324×10^7	square meters
	36.	square miles
	3.68641×10^6	square perches
	3.68641×10^6	square rods
	1.11514×10^8	square yards
trikes	3.	revolutions
tuns	8.	barrels (31.5 U.S. gallons)
	252.	gallons (U.S., liquid)
	953.92	liters

MULTIPLY	BY	TO OBTAIN
varas (California)	33.	inches
varas (Texas)	2.7778	feet
	33.333	inches
	0.846667	meters
	0.9259	yards
velocity of light	9.837×10^8	feet/sec
	1.080×10^9	kilometers/hr
	1.80×10^7	kilometers/min
	2.9979×10^8	meters/sec
	6.706×10^8	miles/hr
	1.118×10^7	miles/min
volts (absolute)	1.0×10^8	abvolts
	0.0033356	statvolts
	0.99967	volts (int'l)
volts (int'l)	1.00033×10^8	abvolts
	0.0033367	statvolts
	1.00033	volts (absolute)
volt-sec	100,000.	kilolines
	1.0×10^8	lines
	1.0×10^8	maxwells
	100.	megalines
	1.	webers
volt-sec/amp	1.	henrys

PROFESSIONAL PUBLICATIONS, INC. ● Belmont, CA

MULTIPLY	BY	TO OBTAIN
watts	3.412	BTU/hr
	0.05683	BTU/min
	9.4871×10^{-4}	BTU/sec
	1.0×10^{7}	dyne-cm/sec
	1.0×10^{7}	ergs/sec
	2655.	foot-lbf/hr
	44.254	foot-lbf/min
	0.73756	foot-lbf/sec
	10,197.	gram-cm/sec
	0.001360	horsepower (metric)
	0.001341	horsepower (U.S.)
	1.	joules/sec
	0.01433	kilogram-cal/min
	2.3885×10^{-4}	kilogram-cal/sec
	0.10197	kilogram (force)-m/sec
	0.001	kilowatts
	668.45	lumens
	2.844×10^{-4}	tons (refrigeration)
watts/cm^2	76,081.	BTU/ft^2-day
	3170.	BTU/ft^2-hr
	859.7	gram-cal/hr-cm^2
	0.2388	gram-cal/sec-cm^2
	10,000.	watts/m^2
watts/cm^2-°C	57.79	BTU
	42,267.	BTU/ft^2-day-°F
	1761.	BTU/ft^2-hr-°F
	859.7	gram-cal/cm^2-hr-°C
	0.2388	gram-cal/cm^2-sec-°C
watts/in^2	8.20	BTU/ft^2-min
	6373.	foot-lbf/ft^2-min
	0.193	horsepower (U.S.)/ft^2
watts/in^2-°F	11,765.	BTU/ft^2-day-°F
	490.2	BTU/ft^2-hr-°F
	239.4	gram-cal/cm^2-hr-°C
	0.06647	gram-cal/cm^2-sec-°C
	0.19313	horsepower (U.S.)/ft^2-°F
	2393.	kilogram-cal/m^2-hr-°C
	0.27833	watts/cm^2-°C

MULTIPLY	BY	TO OBTAIN
watt-cm/cm^2-°C	57.79	BTU-ft/ft^2-hr-°F
	16,641.	BTU-in/ft^2-day-°F
	693.5	BTU-in/ft^2-hr-°F
	859.7	gram-cal-cm/cm^2-hr-°C
	0.2388	gram-cal-cm/cm^2-sec-°C
watt-hr	3.410	BTU
	35,529.	cubic centimeter-atm
	1.255	cubic foot-atm
	3.6×10^{10}	dyne-cm
	3.6×10^{10}	ergs
	2655.2	foot-lbf
	85,429.	foot-poundals
	860.	gram-cal
	3.671×10^7	gram-cm
	0.001360	horsepower-hr (metric)
	0.001341	horsepower-hr (U.S.)
	3600.	joules
	0.859	kilogram-cal
	367.1	kilogram-m
	0.001	kilowatt-hr
	35.53	liter-atm
	36,000.	megalergs
	3600.	watt-sec
watt-sec	9.471×10^{-4}	BTU
	9.869	cubic centimeter-atm
	3.485×10^{-4}	cubic foot-atm
	1.0×10^7	dyne-cm
	0.7376	foot-lbf
	23.730	foot-poundals
	0.23888	gram-cal
	10,197.	gram-cm
	3.78×10^{-7}	horsepower-hr (metric)
	3.725×10^{-7}	horsepower-hr (U.S.)
	1.	joules
	2.39×10^{-4}	kilogram-cal
	0.10197	kilogram-m
	2.778×10^{-7}	kilowatt-hr
	0.001	kilowatt-sec
	0.009869	liter-atm
	10.	megalergs
	2.778×10^{-4}	watt-hr
wavelengths red line cadmium	6438.47	angstroms
	2.11236×10^{-6}	feet
	2.5348×10^{-5}	inches
	6.43847×10^{-7}	meters

MULTIPLY	BY	TO OBTAIN
webers	100,000.	kilolines
	1.0×10^8	lines
	1.0×10^8	maxwells
	100.	megalines
	1.	volt-sec
webers/ampere	1.	henrys
webers/cm^2	1.0×10^8	gauss
	1.0×10^8	lines/cm^2
	6.4516×10^8	lines/in^2
	1.0×10^8	maxwells/cm^2
	6.4516×10^8	maxwells/in^2
	1.0×10^4	teslas
webers/in^2	1.55×10^7	gauss
	1.55×10^7	lines/cm^2
	1.0×10^8	lines/in^2
	1.55×10^7	maxwells/cm^2
	1.0×10^8	maxwells/in^2
	1549.9	teslas
webers/m^2	10,000.	gauss
	10,000.	lines/cm^2
	64,516.	lines/in^2
	10,000.	maxwells/cm^2
	64,516.	maxwells/in^2
	1.	teslas
weeks	7.0192	days (sidereal)
	7.	days (solar)
	0.5	fortnights
	168.	hours (solar)
	10,080.	minutes (solar)
	0.23704	months (lunar)
	0.23014	months (mean)
	606,456.	seconds (sidereal)
	604,800.	seconds (solar)
	0.019126	years (leap)
	0.0191646	years (sidereal)
	0.0191654	years (solar)

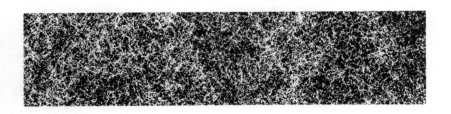

PROFESSIONAL PUBLICATIONS, INC. ● Belmont, CA

MULTIPLY	BY	TO OBTAIN
X-units	0.001	angstroms
	3.28747×10^{-13}	feet
	3.9450×10^{-14}	inches
	1.0×10^{-13}	meters

PROFESSIONAL PUBLICATIONS, INC. ● Belmont, CA

MULTIPLY	BY	TO OBTAIN
yards	0.0250	bolts
	0.00416668	cable lengths
	91.4402	centimeters
	0.0454546	chains (Gunter's or surveyor's)
	0.03	chains (Ramsden's or engineer's)
	2.	cubits
	0.5	fathoms
	3.	feet
	0.00454546	furlongs
	9.	hands
	36.	inches
	9.1440×10^{-4}	kilometers
	1.8939×10^{-4}	leagues (statute)
	4.54546	links (Gunter's or surveyor's)
	3.	links (Ramsden's or engineer's)
	0.914402	meters
	4.93422×10^{-4}	miles (nautical)
	5.68183×10^{-4}	miles (statute)
	1.2	paces
	0.181818	rods (perches or poles)
	0.00833335	skeins
yard^2	(see square yards)	
yard^3	(see cubic yards)	
years	0.01	centuries
	0.1	decades
	0.001	millennia
years (leap)	367.01	days (sidereal)
	366.	days (solar)
	8784.	hours (solar)
	527,040.	minutes (solar)
	12.394	months (lunar)
	12.033	months (mean)
	3.1709×10^7	seconds (sidereal)
	3.1622×10^7	seconds (solar)
	52.286	weeks
	1.00204	years (sidereal)
	1.00208	years (solar)

MULTIPLY	BY	TO OBTAIN
years (sidereal)	366.256	days (sidereal)
	365.256	days (solar)
	8766.15	hours (solar)
	525,969.	minutes (solar)
	12.369	months (lunar)
	12.008	months (mean)
	3.1645×10^7	seconds (sidereal)
	3.1558×10^7	seconds (solar)
	52.180	weeks
	0.997967	years (leap)
	1.00004	years (solar)
years (solar)	366.242	days (sidereal)
	365.242	days (solar)
	8765.8	hours (solar)
	525,949.	minutes (solar)
	12.368	months (lunar)
	12.008	months (mean)
	3.1643×10^7	seconds (sidereal)
	3.1557×10^7	seconds (solar)
	52.178	weeks
	0.99793	years (leap)
	0.99996	years (sidereal)
youngs	647.8	lumens
yukawas	1.0×10^{-5}	angstroms
	3.28084×10^{-15}	feet
	1.	fermis
	3.9370×10^{-14}	inches
	1.0×10^{-15}	meters

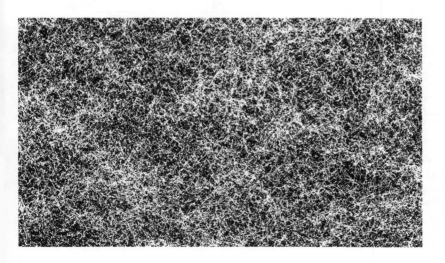

PROFESSIONAL PUBLICATIONS, INC. • Belmont, CA

Temperature Conversions
(Fahrenheit and Celsius, 0° to 100°)

0° to 50°			51° to 100°		
C	C or F	F	C	C or F	F
−17.8	0	32.0			
−17.2	1	33.8	10.6	51	123.8
−16.7	2	35.6	11.1	52	125.6
−16.1	3	37.4	11.7	53	127.4
−15.6	4	39.2	12.2	54	129.2
−15.0	5	41.0	12.8	55	131.0
−14.4	6	42.8	13.3	56	132.8
−13.9	7	44.6	13.9	57	134.6
−13.3	8	46.4	14.4	58	136.4
−12.8	9	48.2	15.0	59	138.2
−12.2	10	50.0	15.6	60	140.0
−11.7	11	51.8	16.1	61	141.8
−11.1	12	53.6	16.7	62	143.6
−10.6	13	55.4	17.2	63	145.4
−10.0	14	57.2	17.8	64	147.2
− 9.4	15	59.0	18.3	65	149.0
− 8.9	16	60.8	18.9	66	150.8
− 8.3	17	62.6	19.4	67	152.6
− 7.8	18	64.4	20.0	68	154.4
− 7.2	19	66.2	20.6	69	156.2
− 6.7	20	68.0	21.1	70	158.0
− 6.1	21	69.8	21.7	71	159.8
− 5.6	22	71.6	22.2	72	161.6
− 5.0	23	73.4	22.8	73	163.4
− 4.4	24	75.2	23.3	74	165.2
− 3.9	25	77.0	23.9	75	167.0
− 3.3	26	78.8	24.4	76	168.8
− 2.8	27	80.6	25.0	77	170.6
− 2.2	28	82.4	25.6	78	172.4
− 1.7	29	84.2	26.1	79	174.2
− 1.1	30	86.0	26.7	80	176.0
− 0.6	31	87.8	27.2	81	177.8
0.0	32	89.6	27.8	82	179.6
0.6	33	91.4	28.3	83	181.4
1.1	34	93.2	28.9	84	183.2
1.7	35	95.0	29.4	85	185.0
2.2	36	96.8	30.0	86	186.8
2.8	37	98.6	30.6	87	188.6
3.3	38	100.4	31.1	88	190.4
3.9	39	102.2	31.7	89	192.2
4.4	40	104.0	32.2	90	194.0
5.0	41	105.8	32.8	91	195.8
5.6	42	107.6	33.3	92	197.6
6.1	43	109.4	33.9	93	199.4
6.7	44	111.2	34.4	94	201.2
7.2	45	113.0	35.0	95	203.0
7.8	46	114.8	35.6	96	204.8
8.3	47	116.6	36.1	97	206.6
8.9	48	118.4	36.7	98	208.4
9.4	49	120.2	37.2	99	210.2
10.0	50	122.0	37.8	100	212.0

Locate temperature in middle column. If in degrees Celsius, read Fahrenheit equivalent in right-hand column; if in degrees Fahrenheit, read Celsius equivalent in left-hand column.

PROFESSIONAL PUBLICATIONS, INC. ● Belmont, CA

Temperature Conversions

original quantity	conversion	new quantity
°Celsius	$\frac{9}{5}$°C + 32	°Fahrenheit
	°C + 273.15	°Kelvin
	$\frac{9}{5}$°C + 491.67	°Rankine
Δ°Celsius	$\frac{9}{5}$Δ°C	Δ°Fahrenheit
	1.0 Δ°C	Δ°Kelvin
	$\frac{9}{5}$Δ°C	Δ°Rankine
°Fahrenheit	$\frac{5}{9}$(°F − 32)	°Celsius
	$\frac{5}{9}$°F + 255.37	°Kelvin
	°F + 459.67	°Rankine
Δ°Fahrenheit	$\frac{5}{9}$Δ°F	Δ°Celsius
	$\frac{5}{9}$Δ°F	Δ°Kelvin
	1.0 Δ°F	Δ°Rankine
°Kelvin	°K − 273.15	°Celsius
	$\frac{9}{5}$(°K − 255.37)	°Fahrenheit
	$\frac{9}{5}$°K	°Rankine
Δ°Kelvin	1.0 Δ°K	Δ°Celsius
	$\frac{9}{5}$Δ°K	Δ°Fahrenheit
	$\frac{9}{5}$Δ°F	Δ°Rankine
°Rankine	$\frac{5}{9}$(°R − 491.67)	°Celsius
	°F − 459.67	°Fahrenheit
	$\frac{5}{9}$°R	°Kelvin
Δ°Rankine	$\frac{5}{9}$Δ°R	Δ°Celsius
	1.0 Δ°R	Δ°Fahrenheit
	$\frac{5}{9}$Δ°R	Δ°Kelvin

PROFESSIONAL PUBLICATIONS, INC. • Belmont, CA

The Elements (Alphabetical According to Symbol)

symbol	name	atomic number	symbol	name	atomic number
Ac	actinium	18	Mn	manganese	25
Ag	silver	47	Mo	molybdenum	42
Al	aluminum	13	Mv	mendelevium	101
Am	americium	95	N	nitrogen	7
Ar	argon[a]	18	Na	sodium	11
As	arsenic	33	Nb	niobium[b]	41
At	astatine	85	Nd	neodymium	60
Au	gold	79	Ne	neon	10
B	boron	5	Ni	nickel	28
Ba	barium	56	No	nobelium	102
Be	beryllium	4	Np	neptunium	93
Bi	bismuth	83	O	oxygen	8
Bk	berkelium	97	Os	osmium	76
Br	bromine	35	P	phosphorus	15
C	carbon	6	Pa	protactinium	91
Ca	calcium	20	Pb	lead	82
Cd	cadmium	48	Pd	palladium	46
Ce	cerium	58	Pm	promethium	61
Cf	californium	98	Po	polonium	84
Cl	chlorine	17	Pr	praseodymium	59
Cm	curium	96	Pt	platinum	78
Co	cobalt	27	Pu	plutonium	94
Cr	chromium	24	Ra	radium	88
Cs	cesium	55	Rb	rubidium	37
Cu	copper	29	Re	rhenium	75
Dy	dysprosium	66	Rf	rutherfordium	104
Es	einsteinium	99	Rh	rhodium	45
Er	erbium	68	Rn	radon[b]	86
Eu	europium	63	Ru	ruthenium	44
F	fluorine	9	S	sulfur	16
Fe	iron	26	Sb	antimony	51
Fm	fermium	100	Sc	scandium	21
Fr	francium	87	Se	selenium	34
Ga	gallium	31	Si	silicon	14
Gd	gadolinium	64	Sm	samarium	62
Ge	germanium	32	Sn	tin	50
H	hydrogen	1	Sr	strontium	38
Ha	hahnium	105	Ta	tantalum	73
He	helium	2	Tb	terbium	65
Hf	hafnium	72	Tc	technetium	43
Hg	mercury	80	Te	tellurium	52
Ho	holmium	67	Th	thorium	90
I	iodine	53	Ti	titanium	22
In	indium	49	Tl	thallium	81
Ir	iridium	77	Tm	thulium	69
K	potassium	19	U	uranium	92
Kr	krypton	36	V	vanadium	23
La	lanthanum	57	W	tungsten[b]	74
Li	lithium	3	Xe	xenon	54
Lr	lawrencium	103	Y	yttrium	39
Lu	lutetium	71	Yb	ytterbium	70
Mg	magnesium	12	Zn	zinc	30
			Zr	zirconium	40

[a] In older literature, symbol A is sometimes used instead of Ar.

[b] Niobium is also known as columbium; radon is also known as emanation (EM); tungsten is also known as wolfram.

Fundamental and Derived Constants

constant	symbol	value[a]
speed of light	c	3.00×10^8 meters/second
permeability constant	μ_0	1.26×10^{-6} henry/meter
permittivity constant	ϵ_0	8.85×10^{-12} farad/meter
elementary charge	e	1.60×10^{-19} coulomb
Avogadro constant	N_0	6.02×10^{23}/mol
electron rest mass	m_e	9.11×10^{-31} kilogram
proton rest mass	m_p	1.67×10^{-27} kilogram
neutron rest mass	m_n	1.67×10^{-27} kilogram
Faraday constant	F	9.65×10^4 coulomb/mol
Planck constant	h	6.63×10^{-34} joule·second
fine structure constant	α	7.30×10^{-3}
electron charge/mass ratio	e/m_e	1.76×10^{11} coulomb/kg
quantum/charge ratio	h/e	4.14×10^{-15} joule·second/coulomb
electron Compton wavelength	λ_C	2.43×10^{-12} meter
proton Compton wavelength	λ_{Cp}	1.32×10^{-15} meter
Rydberg constant	R_∞	1.10×10^7/meter
Bohr radius	a_0	5.29×10^{-11} meter
Bohr magneton	μ_B	9.27×10^{-24} joule/tesla[b]
nuclear magneton	μ_N	5.05×10^{-27} joule/tesla[b]
proton magnetic moment	μ_p	1.41×10^{-26} joule/tesla[b]
universal gas constant	R	8.31 joule/K mol
standard volume of ideal gas	—	2.24×10^{-2} meter^3/mol
Boltzmann constant	k	1.38×10^{-23} joule/K
first radiation constant $\pi\,2\,hc^2$	c_1	3.74×10^{-16} watt/meter^2
second radiation constant hc/k	c_2	1.44×10^{-2} meter/K
Wien displacement constant	b	2.90×10^{-3} meter·K
Stefan-Boltzmann constant	σ	5.67×10^{-8} watt/meter^2·K^4
gravitational constant	G	6.67×10^{-11} newton·meter^2/kg^2

[a] Moles are gram-mol. Multiply values by 1000 to get quantities per kmol.

[b] Tesla = weber/meter^2

Astronomical Data

	sun	earth	moon
mass	1.99×10^{30} kg	5.98×10^{24} kg	7.36×10^{22} kg
radius[c]	6.96×10^8 m	6.38×10^6 m	1.74×10^6 m
density	1460 kg/m^3	5500 kg/m^3	3340 kg/m^3
period of rotation	2.2×10^6 s	8.62×10^4 s	2.36×10^6 s
radius of orbit[c]		1.49×10^{11} m	3.84×10^8 m
period of revolution		3.16×10^7 s	2.36×10^6 s

[c] average value

Greek Alphabet

A	α	alpha		N	ν	nu
B	β	beta		Ξ	ξ	xi
Γ	γ	gamma		O	o	omicron
Δ	δ	delta		Π	π	pi
E	ϵ	epsilon		P	ρ	rho
Z	ς	zeta		Σ	σ	sigma
H	η	eta		T	τ	tau
Θ	θ	theta		Υ	υ	upsilon
I	ι	iota		Φ	ϕ	phi
K	κ	kappa		X	χ	chi
Λ	λ	lambda		Ψ	ψ	psi
M	μ	mu		Ω	ω	omega

SI Prefixes

prefix	symbol	value
exa	E	$\times 10^{18}$
peta	P	$\times 10^{15}$
tera	T	$\times 10^{12}$
giga	G	$\times 10^{9}$
mega	M	$\times 10^{6}$
kilo	k	$\times 10^{3}$
hecto	h	$\times 10^{2}$
deca	da	$\times 10^{1}$
deci	d	$\times 10^{-1}$
centi	c	$\times 10^{-2}$
milli	m	$\times 10^{-3}$
micro	μ	$\times 10^{-6}$
nano	n	$\times 10^{-9}$
pico	p	$\times 10^{-12}$
femto	f	$\times 10^{-15}$
atto	a	$\times 10^{-18}$

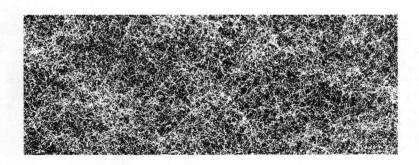

PROFESSIONAL PUBLICATIONS, INC. • Belmont, CA